Advanced Sciences and Technologies for Security Applications

Indexed by SCOPUS

The series Advanced Sciences and Technologies for Security Applications comprises interdisciplinary research covering the theory, foundations and domain-specific topics pertaining to security. Publications within the series are peer-reviewed monographs and edited works in the areas of:

- biological and chemical threat recognition and detection (e.g., biosensors, aerosols, forensics)
- crisis and disaster management
- terrorism
- cyber security and secure information systems (e.g., encryption, optical and photonic systems)
- traditional and non-traditional security
- energy, food and resource security
- economic security and securitization (including associated infrastructures)
- transnational crime
- human security and health security
- social, political and psychological aspects of security
- recognition and identification (e.g., optical imaging, biometrics, authentication and verification)
- smart surveillance systems
- applications of theoretical frameworks and methodologies (e.g., grounded theory, complexity, network sciences, modelling and simulation)

Together, the high-quality contributions to this series provide a cross-disciplinary overview of forefront research endeavours aiming to make the world a safer place.

The editors encourage prospective authors to correspond with them in advance of submitting a manuscript. Submission of manuscripts should be made to the Editor-in-Chief or one of the Editors.

More information about this series at http://www.springer.com/series/5540

Bryanna Fox • Joan A. Reid • Anthony J. Masys
Editors

Science Informed Policing

Springer

Editors
Bryanna Fox
Department of Criminology
University of South Florida
Tampa, FL, USA

Joan A. Reid
Department of Criminology
University of South Florida-St. Petersburg
St. Petersburg, FL, USA

Anthony J. Masys
College of Public Health
University of South Florida
Tampa, FL, USA

ISSN 1613-5113 ISSN 2363-9466 (electronic)
Advanced Sciences and Technologies for Security Applications
ISBN 978-3-030-41286-9 ISBN 978-3-030-41287-6 (eBook)
https://doi.org/10.1007/978-3-030-41287-6

This Springer imprint is published by the registered company Springer Nature Switzerland AG
The registered company address is: Gewerbestrasse 11, 6330 Cham, Switzerland

Introduction

The current policing landscape has seen the rise in serious and organized crime across the globe. Criminals are innovating in real-time, leveraging cyber, social media, enhanced surveillance to support their activities. In so doing, the criminal landscape has become transnational, whereby collaborative networks have flourished, creating greater complexity and novel threats for the international policing community. For example, with the onset of cybercrime as an evolving form of transnational crime, our previous understanding of forensics and criminal business models is challenged by the opaque nature of this criminal domain.

As new threats to local, regional, national, and global security are emerging, leveraging science and technology innovations has become more important. Advances in big data analytics, cyber forensics, surveillance, modeling, and simulation have led to a more data-driven, hypothesis-generated, and model-informed approach to policing, crime prevention, and understanding and deterring offenders. Novel science and technology innovations are presented here to provide insights and pathways that challenge the emerging and complex criminal threat landscape by supporting policing operations.

This edited book explores science and technology innovation and its applications to the policing community. It is divided into two parts:

1. Policing and Crime Prevention
2. Policing Tools and Strategies

Policing and Crime Prevention

Jaynes and Loughran in their chapter "How Offender Decision-Making Can Inform Policing: A Focus on the Perceived Certainty of Apprehension" argue that the perceived certainty of apprehension is a far more effective deterrent than the severity of sanctioning. In concordance with this, many policing strategies have focused on increasing the certainty of apprehension as a key tactic in crime reduction. This chapter describes how recent advances in criminological understanding of perceived certainty are thought to influence offender decision-making. They then illustrate how these findings may inform policing and suggest potential avenues for collaborations between researchers and practitioners to further enhance understanding of offender decision-making and guide evidence-based policing.

Finnegan and Masys in their chapter "An Epidemiological Framework for Investigating Organized Crime and Terrorist Networks" argue that epidemiology, the science of public health, provides a unique approach that can be leveraged and tailored to support policing operations. The systems lens is key to the application of epidemiological models, and hence, the Cynefin framework is introduced to help frame the policing problem space in terms of complexity. This chapter presents a conceptual model that illustrates and operationalizes epidemiological practices to support policing operations.

Chopin and Beauregard in their chapter "Sex Offenders' Forensic Awareness Strategies to Avoid Police Detection" argue that one of the most important costs for offenders is to be identified and arrested by police. In order to avoid arrest, offenders have to rely on specific strategies for "success". In sexual assaults, in addition to appropriate target selection, offenders must make sure to avoid police detection by protecting their identity and cleaning up or destroying forensic evidence that may be left at the crime scene and that can directly lead to their identification. This behavior is known as "forensic awareness." Research on forensic awareness strategies during the crime-commission process is scant. A number of studies looking specifically at sexual crimes identified certain strategies related to forensic awareness. According to the rational choice perspective, the use of strategies to avoid police detection should have an impact on the crime solving by police. Few studies have investigated this issue, and findings are quite unexpected. The aim of this chapter is to review some of the most important empirical knowledge on the use of forensic awareness strategies in sexual crimes.

Shortland and Forest in their chapter "Tracking Terrorism: The Role of Technology in Risk Assessment and Monitoring of Terrorist Offenders" present how technology is increasingly used to help expand the amount of information collected about potential terrorist threats. However, it is clear that an equally important (if not more important) part of these efforts is the ability to separate, within the mass of individuals (or data) identified, those individuals who will conduct acts of terrorism (the minority) from those who will not (the majority). Thus, a core challenge is how information is handled, interpreted, analyzed, and used to inform decisions. In this chapter, they discuss the dynamic interplay between the collection of intelligence (including technology-assisted surveillance) and decision-making. Specifically, they focus on how issues in our understanding of "the terrorist" interact with and impact the use of technology within counter-terrorism. In addition to this, and in support of future research in this area, they highlight some innovative areas for growth and new avenues to facilitate the integration of technology within counter-terrorism.

Reid and Fox in their chapter "Human Trafficking and the Darknet" describe how in an era of unprecedented technological accessibility, coupled with the possibility of online anonymity, human traffickers often operate in the shadows – avoiding interference by typical societal safeguards and law enforcement. This chapter highlights challenges to combatting human trafficking due to the widespread use of mobile technology and the anonymity of the Darknet. The chapter reviews criminal justice resources devoted to combatting human trafficking facilitated by the Internet.

Policing Tools and Strategies

Borum in his chapter "Scientific and Technological Advances in Law Enforcement Intelligence Analysis" describes how a range of professions and business enterprises have moved toward a more science-driven approach to operations. Law enforcement has been no exception. In fact, the modern-day idea of intelligence-led policing (ILP) emerged in the UK in the 1990s as the country was pushing all government services to operate on more of a data-informed, business process or managerial model. This trend led to the development of a British "National Intelligence Model" (NIM), which by 2002, was formally adopted as policy for law enforcement agencies nationwide. ILP has surged in popularity among US law enforcement agencies, although what exactly ILP means in an operational sense and how it is implemented vary considerably. This chapter will not focus primarily on the programmatic aspects of ILP but on scientific and technological advances that have enhanced and accelerated the intelligence analysis process for policing applications.

Houck in his chapter "Improving Criminal Investigations with Structured Analytic Techniques" discusses how the intelligence community has used structured analytic techniques, which are methods designed to reduce bias and increase transparency of process, for years. The techniques force analysts out of routine thinking and away from heuristic habits in order to increase creativity, more comprehensively evaluate the questions, and create a document trail that reveals the thinking process that led to the intelligence product. These methods can be adapted for use in criminal investigations to help reduce bias, improve accuracy, and avoid both wrongful convictions (over 2200 to date) and reparations (more than $2.2 billion) while optimizing resources. The methods shift the investigator from intuitive, daily thinking (System 1 in Kahneman's terminology) to a more analytical approach (System 2) that creates a transparent process, regardless of the outcome. Structured analytical techniques are simple to use, inexpensive, and largely visual; they promote transparency, creativity, and group discussion, leading to better-supported results.

Houck in his chapter "Front-End Forensics: An Integrated Forensic Intelligence Model" describes how forensic science has the ability and capacity to be an active participant in investigations, rather than its traditional passive role. Too often, the evidence is collected and submitted to the laboratory, and the investigators wait for results, creating a lag between when the information is needed and when it is made available. Obstacles to forensic service providers being actively involved in investigations include organizational cultural norms, operational resources, sworn and civilian human resource issues, and media-driven expectations of outcomes. While traditional intelligence analysis is forecasting, forensic science and criminal investigations are reconstructive, and this alters which methods can be and how to improve investigations. What has been characterized as forensic intelligence in the literature is largely forensic support to crime analysis. The benefits to using actual forensic intelligence actively in investigations by shifting forensic results, even if preliminary, to the front-end of the criminal justice system are reduced wrongful arrests and convictions, more efficient use of policing resources, and stronger cases for adjudication. An integrated forensic intelligence model (IFIM) is offered as a roadmap to creating a sustained professional culture of forensic intelligence.

Jeanis in her chapter "Missing Persons and Runaway Youth: The Role of Social Media as an Alert System and Crime Control Tool" argues that missing persons cases have a unique relationship with the media, where increased exposure is said to be a key factor in the recovery of victims and closure of cases. In an attempt to increase exposure for certain types of missing persons cases, news media outlets share case content and alert systems were created; however, the current systems in place have notable limitations. In an attempt to adapt to the proliferation of social media in the lives of everyday Americans, law enforcement agencies have begun to rely on social media as a crime control tool and as a means to raise awareness of missing persons cases. This chapter serves as an overview of the current academic standing of social media as a mechanism of crime control, with a specific focus on how law enforcement can optimize their use of social media in the form of a missing persons outreach tool.

Farrell, Wills, and Nicolas in their chapter "Police Engagement in Multidisciplinary Team Approaches to Commercial Sexual Exploitation of Children" describe how over the past two decades law enforcement has been given new mandates to identify and respond to the commercial sexual exploitation of children (CSEC). While CSEC has been identified as an important community safety issue and the negative impacts on young people are well documented, it has proven to be difficult for the police to identify young people who are at risk of victimization and challenging to develop the trust necessary to gather information needed to support investigations, arrest perpetrators, and provide safety to child victims. In response to these challenges, law enforcement is increasingly partnering with child welfare and service provider stakeholders to improve responses to CSEC in local communities. This chapter describes the problem of CSEC and outlines key aspects of a multidisciplinary team approach to the problem. Using qualitative interview data from participants in six multidisciplinary teams in Massachusetts dedicated to responding to CSEC, they discuss key benefits and challenges to a multidisciplinary team approach to CSEC.

Bennett in his chapter "The Lived Reality of Police Helicopter Operations: Frank and Revealing Interviews with National Police Air Service Personnel" reveals the lived reality of police helicopter operations in England and Wales, from witnessing acts of extreme violence (such as a beheading in North London) to searching for the remains of bodies atomized by locomotive impacts. Highly professional in their outlook and behavior, the men and women of the National Police Air Service

(NPAS) push themselves and their equipment to the limit in their quest to safeguard the public and apprehend criminals. Interviewees claimed a link between teamwork training and performance. Teamwork training – known in aviation as crew resource management – was universally praised. It is recommended that the police service urgently considers the introduction of teamwork training for all officers. The police service is starved of resources. The NPAS needs better equipment – longer-ranged, more capable helicopters with state-of-the-art sensors and communications equipment and protection from small arms fire. The State's primary mission is to ensure the safety and security of its citizens. As a matter of urgency, the government should divert monies from healthcare and education to the police service. Integrity and courage should be recognized and rewarded.

Masys in his chapter "Science Informed Major Event Security Planning: From Vulnerability Analysis to Security Design" describes how the post 9/11 security landscape has seen a fundamental shift in security planning, organization, and management across the national and global security domain. This shift is particularly evident with regard to special events. Major national regional and global events such as the Olympic Games, FIFA World Cup, the US Super Bowl, and the Boston Marathon draw large crowds of spectators and participants, thereby creating vulnerable targets. The 1972 Munich Olympics and 1996 Atlanta Olympics are two examples of pre-9/11 attacks on sporting events. Since 9/11, we have seen attacks on events such as the Boston Marathon in 2013, Las Vegas active shooter in 2017, and the bombing in Manchester at the Ariana Grande concert in 2017. These examples highlight the inherent vulnerabilities and security challenges that reside in major events (sporting, cultural, religious, musical). With this in mind, security budgets have increased to deal with such risks and threats. For example, the security budget for the 2012 London Olympics topped $1.46 billion and the 2010 Vancouver Olympics was over $900 million.

This chapter examines the security challenges and solutions for managing major event security. Scenario planning and disaster forensic methodologies are presented along with the role of predictive analytics in supporting vulnerability analysis and security solution navigation. Key recommendations are made pertaining to vulnerability analysis and the development and design of a security architecture planning framework rooted in a high reliability security organizations.

Contents

Part I
Policing and Crime Prevention

How Offender Decision-making Can Inform Policing: A Focus on the Perceived Certainty of Apprehension

Chae M. Jaynes and Thomas A. Loughran

Abstract Empirical evidence suggests that the perceived certainty of apprehension is a far more effective deterrent than the severity of sanctioning. In concordance with this, many policing strategies have focused on increasing the certainty of apprehension as a key tactic in crime reduction. This chapter describes how recent advances in criminological understanding of perceived certainty are thought to influence offender decision-making. We then illustrate how these findings may inform policing and suggest potential avenues for collaborations between researchers and practitioners to further enhance understanding of offender decision-making and guide evidence-based policing.

Keywords Deterrence · Certainty · Perception updating · Tipping points · Ambiguity

1 Introduction

Research suggests that individuals are sensitive to risk – in that they think about the potential costs of their actions, and likelihood of those costs, in comparison to the potential benefits when making a choice. This is important from a criminal justice perspective because this means that potential offenders are sensitive not only to the rewards they seek to gain from criminal activity, but also the potential costs that may be incurred. Notably, the various benefits of crime may be financial (Loughran et al. 2013; McCarthy and Hagan 2001; Nguyen et al. 2017), but could also include intangible incentives such as earning respect, defending one's honor, or a sense of thrill

C. M. Jaynes (✉)
University of South Florida, Tampa, FL, USA
e-mail: jaynes@usf.edu

T. A. Loughran
Pennsylvania State University, University Park, PA, USA
e-mail: tal47@psu.edu

B. Fox et al. (eds.), *Science Informed Policing*, Advanced Sciences and Technologies for Security Applications,
https://doi.org/10.1007/978-3-030-41287-6_1

(Anderson 1999; Katz 1988; Nagin and Paternoster 1993). Conversely, the potential costs of crime are also numerous and diverse, including risking injury, the loss of esteem among friends and family, pangs of conscience or feelings of shame, and the loss of a valued relationship or job (Braithwaite 1989; Grasmick and Bursik 1990; Williams and Hawkins 1986). But perhaps the most often discussed costs of crime are those which result from interaction with the criminal justice system – referred to as "formal" costs. These costs include criminal justice sanctions such as arrest, fines, drug testing, electronic monitoring, imprisonment, and within some states, execution.

Deterrence theory formally proposes three ways that the criminal justice system can reduce offending: by increasing the swiftness, severity, or certainty of formal sanctioning (Beccaria 1986[1764]). The swiftness tenet (also referred to as "celerity") is the least studied aspect of deterrence (see Pratt and Turanovic 2018 for an overview), and relatedly has only recently began to be deployed as a policy mechanism (see for instance, Project HOPE [Hawken and Kleiman 2009] and 24/7 Sobriety [Kilmer and Midgette 2019]). In contrast, certainty, or what is often operationalized as the probability of apprehension, and severity, typically thought of a harsher/longer sanctions, are the theoretical foundation for many criminal justice policies and practices (Durlauff and Nagin 2011a, b). For instance, any policy which increases the severity of a punishment or the certainty of it happening relies on these principles; well-known examples include the death penalty, three-strike laws, and mandatory arrest.

Given that the severity and certainty principles are at the heart of criminal justice policy, it is of no surprise that they have been subject to vast empirical scrutiny. From this literature, several known findings have emerged. First, there is little empirical support for the claim that more severe sanctions yield a specific deterrent effect. For instance, meta-analyses conducted by Villettaz et al. (2006) and Gendreau and colleagues (1999, 2000, 2001) generally found little support for specific deterrent effects of either custodial (vs. non-custodial) or longer sentences. In fact, these studies suggest that increases in the severity of sanctioning (e.g., length of incarceration) might be criminogenic. Moreover, Nagin et al. (2009) reviewed the literature and called into question the methodological quality of many studies attempting to link longer sentences to recidivism.[1] Second, and in contrast, there is mounting evidence that the certainty of punishment is capable of yielding a deterrent effect. It has been well documented that an increase in perceived certainty is a far better deterrent than an increase in perceived severity (see Nagin 2013; Paternoster 2010 for overviews).

How individuals form beliefs about and respond to risk of apprehension is a rich source of research (Apel 2013; Nagin et al. 2015). Since about the 1970s deterrence scholars have demonstrated that individuals' *perceptions* of punishment are much

---

[1] Nagin et al. (2009:178) note "a key finding of our review is that the great majority of studies point to a null or criminogenic effect of the prison experience on subsequent offending. This reading of the evidence should, at least, caution against wild claims – at times found in "get tough" rhetoric voiced in recent decades – that prisons have special powers to scare offenders straight."

more influential than objective punishment levels (Geerken and Gove 1975). This means that individuals are deterred by what they perceive the certainty of punishment to be, rather than by some true objective likelihood of detection or sanctioning.[2] The implications of these findings are that *perceived certainty* is now the strongest evidence-based tenet of deterrence which can be used as a policy lever to disincentivize criminal behavior.

This is particularly relevant from a policing perspective, as Nagin (2013, p. 206, emphasis not in original) highlighted, "the most important set of actors affecting certainty is the *police*: without detection and apprehension, there is no possibility of conviction or punishment." In concordance with this, many policing strategies have focused on increasing the certainty of apprehension as a key tactic in crime reduction. More directly, Nagin et al. (2015) argued for what they describe as the *sentinel role* of police in reducing both crime and arrests. This chapter will first describe how policing practices have historically sought to influence perceptions of sanction certainty. We will then describe how recent advances in offender decision-making research have improved the understanding of certainty perceptions. Finally, we will describe how these findings may inform policing and suggest potential avenues for collaborations between researchers and practitioners to further enhance and guide evidence-based policing.

2 Deterrence-Based Policing Strategies

2.1 Random Preventative Patrol

During the professional era of policing inspired by O.W. Wilson (1956), police patrolled the streets randomly in patrol cars with the intention of creating an omnipresence that would reduce crime by eliminating perceptions that a crime could be completed successfully. In the 1970s, the Kansas City Police Department collaborated with researchers to empirically test this practice (Kelling et al. 1974). In this experiment, 15 Kansas City beats were randomly assigned to one of three conditions: "reactive" (police responded only to calls for service), "control" (random preventative patrol was maintained as usual), and "proactive" (random preventative patrol was intensified to 2–3 times the usual level). Contrary to the commonly held assumptions at the time, findings established that random preventative patrol within a jurisdiction did not reduce crime. While this finding had significant implications, shifting policing tactics away from this form of patrol, the findings were not intended to mean that the police cannot deter offending or that deterrence does not work.

[2] Moreover, Pogarsky and Loughran (2016) suggested that the idea that there is a single "objective" certainty and severity of punishment is nonsensical because individuals have private information (e.g., criminal expertise and technique) as well as personal attributes (e.g., age, race, gender, prior record) which have known impacts on the likelihood of apprehension as well as the severity of sanctioning.

Rather, they were interpreted to suggest that police time and resources may be more effectively allocated to other tactics and strategies.

2.2 *Increasing the Number of Police*

Increasing police presence is perhaps the most classic deterrence-based approach to reducing crime, through increasing the true likelihood of apprehension as well as the perceived likelihood. Studies evaluating whether increasing the number of police (measured as the number of police per capita) or police resources (measured as police expenditures per capita) reduce crime have produced generally consistent results of an inverse relationship: as police resources increase the crime rate decreases (Marvell and Moody 1996; Levitt 1997, 2002; Evans and Owens 2007, although see McCrary 2002 for an exception). Additionally, research has shown that abrupt changes in police resources, caused by political occupations (Andeneas 1974), budget cuts or police scandals (Heaton 2010; DeAngelo and Hansen 2013; Shi 2009), and terrorist threat levels (e.g, Klick and Tabarrok 2005) are associated with significant increases in crime. This literature, however, has at least two relevant limitations. The first is that the studies are not able to determine whether reductions in crime are the result of increased *deterrent* effects as opposed to incapacitation effects, given that increased policing may result in more individuals arrested and incarcerated (see Durlauf and Nagin 2011a, b for an overview). Additionally, these studies focus primarily on changes to police resources and do not focus on the tactics the police are using while deployed nor how the police influence individual-level perceptions as suggested by deterrence theory.

2.3 *Hot Spots Policing*

The hot spots policing strategy is another example of a deterrence-based policing tactic. This strategy emerged from an empirical observation reported by Sherman et al. (1989) after analyzing a year's worth of spatial data of over 300,000 calls to police from more than 100,000 addresses and intersections in Minneapolis. They found that 50% of calls for police service came from only 3% of places – places which they referred to as crime "hot spots." Weisburd and Green (1995) later found that this unequal geographic distribution of crime was not only unique to Minneapolis, but also appeared in Jersey City, New Jersey with respect to drug crimes (20% of disorder crime and 14% of crimes against persons came from only 56 hot spots); an empirical finding which has since also been observed in Seattle, Washington (Weisburd et al. 2004), Vancouver, Canada (Bratingham and Bratingham 1999), and the Bronx within New York City (Eck et al. 2000).

Armed with the knowledge that crime occurs most frequently in observable microplaces, Sherman and colleagues then developed the deterrence-based hot spot strategy of policing which concentrates police officers in select "hot spot" locations rather than spreading them evenly or randomly across a beat in an effort to increase the certainty of apprehension (Sherman and Weisburd 1995a). In the first randomized controlled trial testing this strategy, Sherman and Weisburd (1995b) found that an increase in patrol presence (an increase of twice the observed patrol presence, on average) in Minneapolis hot spots reduced the total crime calls by 6–13% and observed disorder was reduced by 50%. These promising results prompted considerable academic and practitioner attention and wide-spread efforts to put hot spot policing into practice. In fact, a report produced by the Police Executive Research Forum in 2008 suggested that as many as 90% of United States police agencies used hot spots as a strategy. Rigorous systematic evaluations and meta-analysis of the effects of this strategy have also been carried out, indicating that hot spots policing results in a "small but noteworthy" reduction in crime (Braga et al. 2014). Moreover, there appears to be little-to-no evidence that crime is displaced into nearby locations, but rather, crime control benefits have been found to spill-over or "diffuse" to nearby areas.

2.4 *Problem-Oriented Policing*

The last deterrence-based strategy which has gained momentum is problem-oriented policing. When employing this strategy, police agencies systematically analyze the problems in their jurisdictions, develop solutions, and evaluate the impact of their efforts (National Research Council 2004; National Institute of Justice 2020). While problem-oriented policing can take many forms, perhaps the most well-known example is the Boston Gun Project's Operation Ceasefire (Kennedy et al. 1996, 2001).

The target of Operation Ceasefire was to deter gang violence through increasing the certainty (and severity) of sanctioning if one were to commit serious violent crime. To do this, police collaborated with the Streetworkers (a coalition of social service workers), probation and parole officers, community leadership, and church groups. Together, they identified and reached out to chronic offenders to indicate that violence would not be tolerated. Police made it clear that they had identified the gang members, and if they chose to engage in violent crime, the police and supporting criminal justice professionals and community members would "pull every lever" legally available to them to *ensure* (increasing the perceived certainty of sanctioning) the offender received an intense response (increase the perceived severity of sanctioning). This method sent a clear message that violence would not be tolerated through directly communicating an increased risk to the potential offenders.

Early evaluations of this approach suggested that it was effective in reducing youth homicides, shots-fired calls for service, and gun assaults (Braga et al. 2001). Since its inception, the "pulling levers" form of problem-oriented policing has gained widespread traction and been implemented and evaluated across the country (e.g, Corsaro et al. 2012; McGarrell et al. 2006; Papachristos et al. 2007; Raphael and Ludwig 2003; Tita et al. 2010; Wilson and Chermak 2011). In a recent systematic review, Braga et al. (2018) identified 24 evaluations which met their rigorous criteria (consistent with protocols and conventions of the Campbell Collaboration). They found that focused deterrence strategies such as "pulling levers" are associated with a significant and moderate crime reduction effect. Moreover, research also illustrates that applying this type of problem-orienteered policing strategy to crime hot spots, is more effective than simply increasing police presence in crime hot spots – demonstrating the potential overlap across these deterrence-based strategies (Braga et al. 2014).

3 Recent Innovations in Offender Decision-making Research

While the above efforts provide evidence that deterrence-based policing tactics are associated with crime reduction, recent findings surrounding perceptions of certainty within the offender decision-making literature provide additional insight into how the police may more effectively shape a potential offender's choice. Specifically, these findings highlight the importance of perceptual updating, the potential for tipping points and how ambiguity may be used as a deterrent. Together, this literature has greatly informed the understanding of offender decision-making and may have the potential to enhance policing practices.

3.1 Updating Perceptions

Although research supports that the perceived certainty of apprehension is negatively associated with offending, generally, we know little about where individuals' perceptions of certainty come from or how they are formed. Almost 10 years ago, Paternoster (2010) referred to this as the "dirty little secret in deterrence research" (p. 808), because very few studies had directly examined the mechanism through which the police (or any external policy, organization, experiences, or individuals) are able to alter perceptions. This left many unanswered questions with potentially large-scale policy relevance. For instance, how much does news of a police crackdown really alter a potential offender's perceptions that he or she will be apprehended for committing a crime? If an offender were to commit a crime and get caught, how would this change a certainty perception? Similarly, if the offender were to commit a crime and get away with it, would this alter the certainty perception?

In an early study, Nagin and Paternoster (1993) sought to answer some of these questions. Specifically, among a convenience sample of students, they found that individuals were less likely to express a willingness to drink and drive if they had been informed of a police crackdown on drunk driving relative to a state police budget cutback. In addition, the young adults perceived less likelihood of detection if they had shorter distances to drive home (1 mile compared to 10 miles) or could travel on back roads relative to busy roads; all of which provided initial evidence that police can reduce crime through influencing individuals' perceptions of certainty.

Following this study, scholars grew increasingly interested in understanding *how* offenders update their perceptions after offending. Notably, this line of research diverged from traditional deterrence scholarship focusing on how perceptions influence the choice to commit crime (perceived certainty ➔ crime) to evaluating how offending outcomes (experiences with punishment or punishment avoidance) influence perceptions (crime ➔ perceived certainty) – a reversal of the causal ordering also referred to as "the experiential effect" (Saltzman et al. 1982; Stafford and Warr 1993). The literature has since not only shown that experiencing an arrest increases an individual's risk perception, but also that those who have avoided punishment have reduced perceptions of risk (Lochner 2007; Matsueda et al. 2006; Paternoster and Piquero 1995; Pogarsky et al. 2004). This finding is notable, as prior work raised questions as to whether an apprehension may actually result in a *decrease* in perceived certainty, given that Piquero and Pogarsky (2003) found that getting apprehended encouraged (rather than discouraged) students' willingness to drink and drive.[3]

Anwar and Loughran (2011) took this scholarship a step further by applying a Bayesian learning model and demonstrating that high-risk individuals who had been involved in serious offending during adolescence alter their perceived certainty of apprehension in response to whether they are caught by the police. Specifically, they found that when an individual commits a crime and is arrested, the arrest will increase his or her perceived probability of arrest by 6.3% compared to having avoided detection. Moreover, the study also found that more experienced offenders (offenders who have committed more offences and thus have more potential apprehension outcomes) update their perceptions of the certainty of arrest less than those who are less experienced offenders. This is important from a policing perspective because it demonstrates that the police may have more of an opportunity to increase perceptions of certainty among relatively nascent offenders and have less potential influence on those who are further into their criminal career.

Wilson et al. (2017) later expanded upon Anwar and Loughran's work to note that individuals not only update their risk perceptions based on their *own*

[3] They explained this unexpected finding as evidence of a resetting effect based on the "gambler's fallacy." This is the idea that when rolling a regular 6-sided dice, if an individual has rolled even three times in a row, the person may think "I have a high likelihood of rolling an odd number next time, because surely an odd number has to come up next!" This is a decision-making bias or "fallacy" because each single roll of a dice is independent with .50 probability or rolling an even and .50 probability or rolling an odd. Notably, the previous roll (because it is independent) has no influence on the outcome of the next roll. However, after rolling an even several times, an individual may feel reassured (or perceive) that they're more likely to roll an even on the next roll. This exact idea was applied to offending. For example, if a person is apprehended for committing a crime – they could cognitively believe events are interdependent and now that they have been apprehended once, they're sure to get away with it next time, creating an "emboldening effect" where the perceived certainty of apprehension actually decreases in response to apprehension.

experiences with arrest or arrest avoidance, but that they update risk perceptions based on vicarious experiences as well. For instance, individuals with family members who committed a crime and avoided apprehension were more likely to decrease their personal perceptions of sanction certainty. These findings suggest that the police are not only able to influence an apprehended offender's perceptions, but that an apprehension can have an even larger dispersed deterrent effect through increasing certainty perceptions among the offender's friends and family.

3.2 Tipping Points and Threshold Effects

In addition to examining how perceptions of certainty are updated, research has also begun to more carefully consider the functional form of the relationship between perceived certainty and offending. More specifically, the certainty ➔ offending relationship had most often been hypothesized (and supported empirically) as being linear – a one-unit increase in certainty would result in *the same* reduction in offending regardless of where an individual's perceived certainty fell in the certainty distribution. However, scholars began to question whether this linear hypothesis made sense theoretically and had validity empirically, because for some time scholars had shown evidence of a "tipping effect" at the aggregate level – a threshold in apprehension probability that has to be met in order for police efforts to have the hypothesized deterrent effect. Tittle and Rowe (1974) first made this suggestion, after finding that the arrest clearance rate (historically considered an objective measure of certainty) was only negatively associated with the crime rate for counties and municipalities in Florida where the clearance rate was above .30. In areas where the crime rate was below .30, there was no significant correlation between the clearance rate and crime rate. Brown (1978) and Chamlin (1991) also reached similar findings, identifying a tipping point around .30–.40 in California and Pennsylvania, respectively, particularly among small cities. Yu and Liska (1993) later extended this research by disaggregating the analysis by crime type and race in order to understand why few large cities reached the .30–.40 level required to observe a strong relationship in prior research. Drawing on national data from the Uniform Crime Reports (UCR) and the National Crime Survey (NCS), they found that racial majorities were less likely to experience certainty levels reaching a tipping point, but that racial minorities did reach these levels, so that a tipping point effect was observed (see also Chilton 1982).

Although a tipping point had been observed at the macro level (e.g, city, municipality), until fairly recently, there was little evidence of a tipping point at the individual level. Loughran et al. (2012) evaluated the relationship between perceived certainty of punishment and self-reported offending among a sample of serious youth offenders. There results made two important contributions. First, they demonstrated that consistent with research at the macro-level, there was significant evidence of a threshold tipping point such that after the certainty of apprehension reached .30–.40, an increase in certainty resulted in a reduction in criminal offending. Second, they found that although there was a deterrent effect at mid-levels of perceived certainty, the deterrent effect was much stronger for individuals at the high-end of the certainty spectrum. This is consistent with the notion that individuals mentally overweight high-probabilities to an extent that they almost consider the likelihood of apprehension to be certain.

These findings have at least two important implications for policing practices. The first is that police tactics targeting crime reductions through increasing the certainty of apprehension may not be effective if the tactic does not increase individuals risk perceptions to a threshold of at least .40. Second, the effect of a change in risk perception on offending is largely contingent on where an individual's certainty perception began (referred to as differential marginal effects). For instance, assume that a change in police tactic, such as a drug crackdown, increased perceptions of apprehension by .05. Based on Loughran et al.'s (2012) findings, this increase would have a much greater deterrent effect for individuals who already perceived the risk of apprehension to be high (beginning at .90 and increasing to .95) than individuals who originally perceived their likelihood of detection to be low (beginning at .40 and increasing to .45). Though both increases would, theoretically, have a significant deterrent effect, it would be much smaller for the group of individuals police efforts most often intend to deter (those who perceive the likelihood of apprehension to be low). Practically, this could mean police efforts need to target larger increases in perceived certainty to observe desirable reductions in crime. Notably, there are few estimates indicating how much a change in police tactic would practically influence an individual's certainty perception.

3.3 Ambiguity

Thus far, the discussion has focused on how increasing perceptions of certainty may influence offending, particularly focusing on the *level* of certainty (e.g., increasing the perceived risk of detection from .20 to .30). However contradictory it may seem at first, research has recently begun to indicate that there may be opportunity to create deterrent effects by making individuals' perceptions more *un*certain, or *ambiguous*. This line of research draws on the fact that the true certainty of apprehension cannot realistically be known to a potential offender. Rather, an individual's perceived likelihood of apprehension is subjective and simply a best guess; and individuals differ in their estimate precision. To provide an example, imagine that two potential drug dealers were asked about their certainty of apprehension. The first, is fairly sure there is a .40 likelihood he will be caught, but admits he is not entirely sure, and the likelihood could really be from .30 to .50. The second, also states that he thinks there is a .40 likelihood of apprehension, but is far less sure of his answer, admitting that the likelihood of apprehension is somewhere between .10 and .70. In both cases, .40 is the expected value, or mean likelihood of apprehension. However, the second potential offender is much less certain of his perception estimate – making his certainty estimate more *uncertain*.

In his review of police crackdowns completed nearly 30 years ago, Sherman (1990) highlighted that in practice the likelihood of apprehension for committing an offense is actually quite low, and that "permanent law enforcement priorities may make the risks of punishment all too predictable for criminals" then suggesting "there is no reason why criminal justice policy has to provide such easy bets for criminals" (p. 7). He acknowledged that the reality of police targeted efforts is that they tend to offer high certainty of a low risk of apprehension, rather than low certainty about the risk of punishment on any given day. Therefore, he surmised that if police could exploit inherent ambiguity, making even a low perceived likelihood of apprehension more uncertain, it may be a more cost-effective strategy compared to attempting to increase the perceived certainty of apprehension level.

Nagin (1998) later expanded on this idea, applying the concept of 'ambiguity aversion" from behavioral decision-making theory. This concept suggests that while two perceptions of certainty (.30–.50 and .10–.70, drawn from the earlier example) may have the same expected value (.40), one estimate has more variance (or ambiguity), and individuals are *averse* to ambiguity in that they prefer gambles with less variance (or more certainty) (Camerer and Weber 1992). While Sherman (1990) had demonstrated that police crackdowns are typically successful in producing a deterrent effect early on (termed "initial deterrence"), there was also evidence of a "initial deterrence decay" as "offenders learn through trial and effort that they had overestimated the certainty of getting caught at the beginning of the crackdown", as well as evidence of "residual deterrence" which is a crime reduction effect that continues beyond the intervention period until offenders learn through trial and error that it is "safe to offend" (p. 10). Nagin (1998) emphasized that individuals' tendency toward ambiguity aversion may be one explanation for these observations, if a crackdown increases the variance (or ambiguity) in an offender's certainty estimates. Additionally, as a crackdown continues, and individuals may become more certain of the likelihood of apprehension (reducing ambiguity), producing less of a deterrence effect.

Loughran et al. (2011) were the first to empirically evaluate whether ambiguity in perceptions of sanction risk influenced the decision to offend. In a sample of high-risk juvenile offenders who had committed serious crime, they found that consistent with the predictions of Sherman (1990) and Nagin (1998), there was significant evidence of ambiguity aversion, where an increase in certainty ambiguity was associated with a decrease in offending, particularly for crimes that involved no contact between offenders and victims.[4] Additionally, they did not simply find blanket ambiguity effects. At lower ends of the risk continuum (<.30), increased ambiguity was a significant deterrent. However, for individuals with high-risk perceptions (>.70), ambiguity was a less effective deterrent. Additionally, for crimes involving contact between an offender and victim, ambiguity had no independent deterrent

[4] No contact crimes included breaking in, stealing, theft, and vandalism. Contact crimes included fighting, stabbing, and robbery with a gun.

effect, and in fact, was found to encourage offending among individuals with low perceptions of detection (<.30).

Two more recent studies on perception formation provide additional context. First, with respect to ambiguity, Pickett et al. (2015) provided experimental results suggesting that uncertainty about subjective beliefs may not necessarily be related to lack of exposure to information. This suggests that ambiguity in perceptions may be more systematic to individuals or situations as opposed to rationally updated. Second, Thomas et al. (2018) study found that individual perceptions about different crimes were largely stable across crimes and within persons, yet otherwise largely not grounded in objective values or comparable across persons, a result they compared to the concept of coherent arbitrariness from behavioral economics (Airley et al. 2003). This suggests that while individual beliefs about risk are perhaps grounded and informative, they are nonetheless more idiosyncratic to the individual than a literal interpretation of rational choice might imply.[5]

Substantively, these findings have several implications. First, they suggest that police could increase the effectiveness of their deterrence efforts by increasing uncertainty in risk perceptions, particularly, for non-contact crimes. Second, the results suggest exploiting ambiguity should not be used as a strategy for reducing contact crimes. Third, increasing ambiguity is most effective among those with already low perceptions of detection (<.30). This is particularly noteworthy as individuals with low risk of detection are least likely to be deterred from a traditional deterrence perspective (low certainty would be associated with higher offending) and based on evidence of a tipping point (certainty of apprehension less than .30 would have no effect).

Practically, the police could increase ambiguity surrounding the certainty of apprehension a few ways. For instance, when policing within crime hot spots, patrols could be randomized – forcing offenders to guess where the police will be and when they will be there. Based on the history of policing, some may question the suggestion of random preventative patrol, as this tactic was not found to be effective years ago (Kelling et al. 1974). However, this suggestion is distinguishable from the random preventative patrol of the past, in that it encourages the use of hot spots policing, making patrol random *within or across* hot spots. Additionally, the random preventative patrol of the past intended to create a deterrent effect primarily through automotive patrol, whereas modern policing now uses evidence-based strategies such as community and problem-oriented policing. In addition, rather than sending police uniformly across hot spots, creating a fairly unambiguous low level (albeit now a higher level because of the crackdown) of detection, police could be sent to a single hotspot or only a few hot spots on a particular day. While the objective certainty of detection would be the same across hot spots on average, and

[5] For instance, as Thomas et al. (2018) describe, two individuals who report the same perception about risk for a certain crime (e.g., 40%) may actually have very different beliefs about what this means, though each would feel that this crime is riskier than a different crime for which they feel the risk is only 20%, even if the absolute differences are not meaningful.

no additional policing resources would to be used, ambiguity may be increased and provide an enhanced deterrent effect.

4 Future Directions

Although findings from the offender decision-making literature discussed above carry implications for policing, there are still many questions left unanswered and limitations surrounding the underlying mechanisms through which police influence offender decision-making. For instance, all (or nearly all) of the research within this chapter falls into one of two categories. The first, are *policing* studies examining the effectiveness of a policing tactic on reducing crime. The second, are *offender decision-making* studies, evaluating how perceptions of certainty influence the choice to offend. Although these two spheres of research have significant overlap, as we have hopefully demonstrated within this chapter, the bodies of research have remained largely separate historically. The former has largely failing to consider the deterrence mechanism underlying observed reductions in offending – individual level changes in perceptions, and the latter has typically failed to focus on external influences on perceptions – how policing tactics can influence individual level perceptions. We believe these bodies of research have remained unnecessarily distinct and each specialty could inform the other. While findings from offender decision-making have implications for policing, these implications (to our knowledge) have yet to be rigorously empirically evaluated. Therefore, we suggest that practitioners and scholars consider evaluating these implied tactics, while also collecting information on individual-level perceptions, to determine whether the deterrence mechanism assumed to be influence crime outcomes is indeed behaving as expected within deterrence theory. In addition, it is our hope that findings surrounding perceptual updating, tipping points, and ambiguity can aid policing practices to create a more effective and efficient system.

Finally, we conclude by noting that there is a clear body of evidence suggesting that certainty is linked to deterrence, and such deterrent effects can typically occur ex ante to a crime commission, which if preventable is less costly to both society (in the form of costly punishment and losses associated with victimization) and the potential offender (in the form of being punished and subsequent difficulties). However, the more fine-grained mechanisms of the certainty effect and how it might be utilized more effectively as a crime prevention tool are still open questions. As such, we advocate for criminologists to look to studies outside the criminological literature which highlight creative ways to study, for instance, exogenous shifts in police levels and the subsequent crime response, and individual decision-making biases which can potentially be exploited by police.

References

Andenaes, J. (1974). *Punishment and deterrence*. Ann Arbor: University of Michigan Press.

Anderson, E. (1999). *The code of the street*. New York: W.W. Norton & Company.

Anwar, S., & Loughran, T. A. (2011). Testing a Bayesian learning theory of deterrence among serious juvenile offenders. *Criminology, 49*(3), 667–698.

Apel, R. (2013). Sanctions, perceptions, and crime: Implications for criminal deterrence. *Journal of Quantitative Criminology, 29*(1), 67–101.

Ariely, D., Loewenstein, G., & Prelec, D. (2003). "Coherent arbitrariness": Stable demand curves without stable preferences. *The Quarterly Journal of Economics, 118*(1), 73–106.

Beccaria, C. (1986[1764]). *On crimes and punishments*. Indianapolis: Hackett Publishing Company.

Braga, A. A., Kennedy, D. M., Waring, E. J., & Piehl, A. M. (2001). Problem-oriented policing, deterrence, and youth violence: An evaluation of Boston's operation ceasefire. *Journal of Research in Crime and Delinquency, 38*(3), 195–225.

Braga, A. A., Papachristos, A. V., & Hureau, D. M. (2014). The effects of hot spots policing on crime: An updated systematic review and meta-analysis. *Justice Quarterly, 31*(4), 633–663.

Braga, A. A., Weisburd, D., & Turchan, B. (2018). Focused deterrence strategies and crime control: An updated systematic review and meta-analysis of the empirical evidence. *Criminology & Public Policy, 17*(1), 205–250.

Braithwaite, J. (1989). *Crime, shame and reintegration*. New York: Cambridge University Press.

Brantingham, P. L., & Brantingham, P. J. (1999). A theoretical model of crime hot spot generation. *Studies on Crime & Crime Prevention, 8*(3), 7–26.

Brown, C. (1978). Deterrence and accident compensation schemes. *University of Western Ontario Law Review, 17*, 111.

Camerer, C. & Weber, M. (1992). Recent developments in modeling preferences: Uncertainty and ambiguity. *Journal of Risk and Uncertainty 5*(4), 325–370.

Chamlin, M. (1991). A longitudinal analysis of the arrest–crime relationship: A further examination of the tipping effect. *Justice Quarterly, 8*(2), 187–199.

Chilton, R. (1982). Analyzing urban crime data: Deterrence and the limitations of arrests per offense ratios. *Criminology, 19*(4), 590–607.

Corsaro, N., Hunt, E. D., Hipple, N. K., & McGarrell, E. F. (2012). The impact of drug market pulling levers policing on neighborhood violence: An evaluation of the high point drug market intervention. *Criminology & Public Policy, 11*(2), 167–199.

DeAngelo, G., & Hansen, B.. (2013). *Life and death in the fast lane: Police enforcement and roadway safety*. Available at SSRN: https://doi.org/10.2139/ssrn.1940134

Durlauf, S., & Nagin, D. (2011a). Imprisonment and crime can both be reduced? *Criminology & Public Policy, 10*(1), 13–54.

Durlauf, S., & Nagin, D. (2011b). The deterrent effect of imprisonment. In P. J. Cook, J. Ludwig, & J. McCrary (Eds.), *Controlling crime: Strategies and tradeoffs* (pp. 43–94). Chicago: University of Chicago Press.

Eck, J. E., Gersh, J. S., & Taylor, C. (2000). Finding crime hot spots through repeat address mapping. In V. Goldsmith, P. G. McGuire, J. B. Mollenkoph, & T. A. Ross (Eds.), *Analyzing crime patterns: Frontiers of practice* (pp. 49–64). Thousand Oaks: Sage.

Evans, W. N., & Owens, E. G. (2007). COPS and crime. *Journal of Public Economics, 91*(1–2), 181–201.

Geerken, M. R., & Gove, W. R. (1975). Deterrence: Some theoretical considerations. *Law and Society Review, 9*(3), 497–513.

Gendreau, P., Goggin, C., & Cullen, F. T. (1999). *The effects of prison sentences on recidivism* (pp. 4–5). Ottawa: Solicitor General Canada.

Gendreau, P., Goggin, C., & Fulton, B. (2000). Intensive probation in probation and parole settings. In C. R. Hollin (Ed.), *Handbook of offender assessment and treatment* (pp. 195–204). Chichester: Wiley.

Gendreau, P., Goggin, C., Cullen, F. T., & Andrews, D. A. (2001). The effects of community sanctions and incarceration on recidivism. In *Compendium of effective correctional programs* (Volume 1, Chapter 4). Ottawa: Correctional Service of Canada, Solicitor General of Canada.

Grasmick, H. G., & Bursik, R. J., Jr. (1990). Conscience, significant others, and rational choice: Extending the deterrence model. *Law and Society Review, 24*(3), 837–861.

Hawken, A., & Kleiman, M. (2009). *Managing drug involved probationers with swift and certain sanctions: Evaluating Hawaii's HOPE*. Washington, DC: National Institute of Justice.

Heaton, P. (2010). Understanding the effects of anti-profiling policies. *The Journal of Law and Economics, 53*(1), 29–64.

Katz, J. (1988). *Seductions of crime*. New York: Basic Books.

Kelling, G. L., Pate, T., Dieckman, D., & Brown, C. E. (1974). *The Kansas City preventive patrol experiment*. Washington, DC: Police Foundation.

Kennedy, D. M., Piehl, A. M., & Braga, A. A. (1996). Youth violence in Boston: Gun markets, serious youth offenders, and a use-reduction strategy. *Law and Contemporary Problems, 59*(1), 147–196.

Kennedy, D. M., Braga, A. A., Piehl, A. M., & Waring, E. J. (2001). *Reducing gun violence: The Boston gun project's operation ceasefire* (Research report). Washington, DC: US National Institute of Justice.

Kilmer, B., & Midgette, G. (2019). *Using certainty and celerity to deter crime* (RAND Social and Economic Well-Being Working Paper WR-1190-1-NIAAA). Santa Monica: RAND. Available at https://doi.org/10.7249/WR1190.

Klick, J., & Tabarrok, A. (2005). Using terror alert levels to estimate the effect of police on crime. *The Journal of Law and Economics, 48*(1), 267–279.

Levitt, S. D. (1997). Using electoral cycles in police hiring to estimate the effect of police on crime. *American Economic Review, 87*(3), 270–290.

Levitt, S. D. (2002). Using electoral cycles in police hiring to estimate the effects of police on crime: Reply. *American Economic Review, 92*(4), 1244–1250.

Lochner, L. (2007). Individual perceptions of the criminal justice system. *American Economic Review, 97*(1), 444–460.

Loughran, T. A., Paternoster, R., Piquero, A. R., & Pogarsky, G. (2011). On ambiguity in perceptions of risk: Implications for criminal decision making and deterrence. *Criminology, 49*(4), 1029–1061.

Loughran, T. A., Pogarsky, G., Piquero, A. R., & Paternoster, R. (2012). Re-examining the functional form of the certainty effect in deterrence theory. *Justice Quarterly, 29*(5), 712–741.

Loughran, T. A., Nguyen, H., Piquero, A. R., & Fagan, J. (2013). The returns to criminal capital. *American Sociological Review, 78*(6), 925–948.

Marvell, T. B., & Moody, C. E. (1996). Specification problems, police levels, and crime rates. *Criminology, 34*(4), 609–646.

Matsueda, R. L., Kreager, D. A., & Huizinga, D. (2006). Deterring delinquents: A rational choice model of theft and violence. *American Sociological Review, 71*(1), 95–122.

McCarthy, B., & Hagan, J. (2001). When crime pays: Capital, competence, and criminal success. *Social Forces, 79*(3), 1035–1060.

McCrary, J. (2002). Using electoral cycles in police hiring to estimate the effect of police on crime: Comment. *American Economic Review, 92*(4), 1236–1243.

McGarrell, E. F., Chermak, S., Wilson, J. M., & Corsaro, N. (2006). Reducing homicide through a "lever-pulling" strategy. *Justice Quarterly, 23*(02), 214–231.

Nagin, D. S. (1998). Criminal deterrence research at the outset of the twenty-first century. *Crime and Justice, 23*, 1–42.

Nagin, D. S. (2013). Deterrence in the twenty-first century. *Crime and Justice, 42*(1), 199–263.

Nagin, D. S., & Paternoster, R. (1993). Enduring individual differences and rational choice theories of crime. *Law and Society Review, 27*(3), 467–496.

Nagin, D. S., Cullen, F. T., & Jonson, C. L. (2009). Imprisonment and reoffending. *Crime and Justice, 38*(1), 115–200.

Nagin, D. S., Solow, R. M., & Lum, C. (2015). Deterrence, criminal opportunities, and police. *Criminology, 53*(1), 74–100.

National Institute of Justice (2020). Problem-Oriented Policing. Retrieved 2020 from https://www.crimesolutions.gov/PracticeDetails.aspx?ID=32

National Research Council (2004). Fairness and effectiveness in policing: The evidence. In W. Skogan & K. Frydl (Eds.). *Committee on law and justice, division of behavioral and social sciences and education* (Committee to Review Research on Police Policy and Practices). Washington, DC: National Academies Press.

Nguyen, H., Loughran, T. A., Paternoster, R., Fagan, J., & Piquero, A. R. (2017). Institutional placement and illegal earnings: Examining the crime school hypothesis. *Journal of Quantitative Criminology, 33*(2), 207–235.

Papachristos, A. V., Meares, T. L., & Fagan, J. (2007). Attention felons: Evaluating project safe neighborhoods in Chicago. *Journal of Empirical Legal Studies, 4*(2), 223–272.

Paternoster, R. (2010). How much do we really know about criminal deterrence. *The Journal of Criminal Law & Criminology, 100*(3), 765–824.

Paternoster, R., & Piquero, A. (1995). Reconceptualizing deterrence: An empirical test of personal and vicarious experiences. *Journal of Research in Crime and Delinquency, 32*(3), 251–286.

Pickett, J. T., Loughran, T. A., & Bushway, S. (2015). On the measurement and properties of ambiguity in probabilistic expectations. *Sociological Methods & Research, 44*(4), 636–676.

Pogarsky, G., & Loughran, T. A. (2016). The policy-to-perceptions link in deterrence: Time to retire the clearance rate. *Criminology & Public Policy, 15*(3), 777–790.

Pogarsky, G., & Piquero, A. R. (2003). Can punishment encourage offending? Investigating the "resetting" effect. *Journal of Research in Crime and Delinquency, 40*(1), 95–120.

Pogarsky, G., Piquero, A. R., & Paternoster, R. (2004). Modeling change in perceptions about sanction threats: The neglected linkage in deterrence theory. *Journal of Quantitative Criminology, 20*(4), 343–369.

Police Executive Research Forum. (2008). *Violent crime in America: What we know about hot spots enforcement*. Washington, DC: Police Executive Research Forum.

Pratt, T. C., & Turanovic, J. J. (2018). Celerity and deterrence. In D. S. Nagin, F. T. Cullen, & C. L. Jonson (Eds.), *Deterrence, choice, and crime: Contemporary perspectives* (Advances in criminological theory) (pp. 187–210). New York: Routledge.

Raphael, S., & Ludwig, J. (2003). Prison sentence enhancements: The case of project exile. In J. Ludwig & P. J. Cook (Eds.), *Evaluating gun policy: Effects on crime and violence* (pp. 251–286). Washington, DC: Brookings Institution Press.

Saltzman, L., Paternoster, R., Waldo, G. P., & Chiricos, T. G. (1982). Deterrent and experiential effects: The problem of causal order in perceptual deterrence research. *Journal of Research in Crime and Delinquency, 19*(2), 172–189.

Sherman, L. W. (1990). Police crackdowns: Initial and residual deterrence. *Crime and Justice, 12*, 1–48.

Sherman, L. W., & Weisburd, D. (1995a). Does patrol prevent crime? The Minneapolis hot spots experiment. In K. Miyazawa & S. Miyazawa (Eds.), *Crime presentation in the urban community*. Boston: Kiuwer.

Sherman, L. W., & Weisburd, D. (1995b). General deterrent effects of police patrol in crime "hot spots": A randomized, controlled trial. *Justice Quarterly, 12*(4), 625–648.

Sherman, L. W., Gartin, P. R., & Buerger, M. E. (1989). Hot spots of predatory crime: Routine activities and the criminology of place. *Criminology, 27*(1), 27–56.

Shi, L. (2009). The limit of oversight in policing: Evidence from the 2001 Cincinnati riot. *Journal of Public Economics, 93*(1–2), 99–113.

Stafford, M. C., & Warr, M. (1993). A reconceptualization of general and specific deterrence. *Journal of Research in Crime and Delinquency, 30*(2), 123–135.
Thomas, K. J., Hamilton, B. C., & Loughran, T. A. (2018). Testing the transitivity of reported risk perceptions: Evidence of coherent arbitrariness. *Criminology, 56*(1), 59–86.
Tita, G., Riley, K. J., Ridgeway, G., Grammich, C. A., Abrahamse, A., & Greenwood, P. W. (2010). *Reducing gun violence: Results from an intervention in East Los Angeles*. Santa Monica: RAND.
Tittle, C. R., & Rowe, A. R. (1974). Certainty of arrest and crime rates: A further test of the deterrence hypothesis. *Social Forces, 52*(4), 455–462.
Villettaz, P., Killias, M., & Zoder, I. (2006). The effects of custodial vs. non-custodial sentences on re-offending: A systematic review of the state of knowledge. *Campbell Systematic Reviews, 2*(1), 1–69.
Weisburd, D., & Green, L. (1995). Policing drug hot spots: The Jersey City drug market analysis experiment. *Justice Quarterly, 12*(4), 711–735.
Weisburd, D., Bushway, S., Lum, C., & Yang, S. M. (2004). Trajectories of crime at places: A longitudinal study of street segments in the city of Seattle. *Criminology, 42*(2), 283–322.
Williams, K. R., & Hawkins, R. (1986). Perceptual research on general deterrence: A critical review. *Law and Society Review, 20*(4), 545–572.
Wilson, O. W. (1956). Basic police policies. *The Police Chief*, November: 28–29.
Wilson, J. M., & Chermak, S. (2011). Community-driven violence reduction programs: Examining Pittsburgh's one vision one life. *Criminology & Public Policy, 10*(4), 993–1027.
Wilson, T., Paternoster, R., & Loughran, T. (2017). Direct and indirect experiential effects in an updating model of deterrence: A research note. *Journal of Research in Crime and Delinquency, 54*(1), 63–77.
Yu, J., & Liska, A. E. (1993). The certainty of punishment: A reference group effect and its functional form. *Criminology, 31*(3), 447–464.

An Epidemiological Framework for Investigating Organized Crime and Terrorist Networks

Jennifer C. Finnegan and Anthony J. Masys

Abstract In the last two decades, organized crime and terrorism has grown more complex, posing evolving challenges to law enforcement nationally, regionally and globally. Within the US '…there is a transformation in how criminals are using technology to invent new types of crime, and are creating new methods for committing traditional crimes' (Wexler C. Critical issues in policing series new national commitment required: The changing nature of crime and criminal investigations police executive research forum, Washington, D.C. 20036 retrieved from https://www.policeforum.org/assets/ChangingNatureofCrime.pdf: 2018). In addition, criminals and terrorists have adopted more agile structural models and internationalized their operations. The UK regards organized crime as a threat to national security which costs the UK at least £32 billion each year, leads to loss of life, and can deprive people of their security and prosperity (https://www.gov.uk/government/publications/official-development-assistance-oda-fco-programme-spend-objectives-2017-to-2018/combatting-serious-and-organised-crime-objectives-2017-to-2018).

Given this threat and risk landscape, it has been argued that the '…science of criminal investigations is changing rapidly, and many law enforcement agencies are not prepared for the changes that are taking place' (Wexler C. Critical issues in policing series new national commitment required: The changing nature of crime and criminal investigations police executive research forum, Washington, D.C. 20036 retrieved from https://www.policeforum.org/assets/ChangingNatureofCrime.pdf: 2018). The inherent complexity associated with serious and organized crime and terrorism calls for intervention and investigative strategies and practices rooted in a science-informed, evidence-based approach. Epidemiology, the science of public health, provides a unique approach that can be leveraged and tailored to support policing operations. The systems lens is key to the application of epidemiological models, and hence the Cynefin framework is introduced to help frame the policing problem space in terms of complexity.

J. C. Finnegan · A. J. Masys (✉)
College of Public Health, University of South Florida, Tampa, FL, USA
e-mail: tmasys@usf.edu

B. Fox et al. (eds.), *Science Informed Policing*, Advanced Sciences and Technologies for Security Applications,
https://doi.org/10.1007/978-3-030-41287-6_2

This paper presents a conceptual model that illustrates and operationalizes epidemiological practices to support policing operations.

Keywords Epidemiology · Organized crime · Terrorism · Cynefin · Systems thinking

1 Introduction

In the last two decades, organized crime and terrorism has grown more complex, posing evolving challenges to law enforcement nationally, regionally, and globally. Within the US '…there is a transformation in how criminals are using technology to invent new types of crime, and are creating new methods for committing traditional crimes' (Wexler 2018). In addition, criminals and terrorists have adopted more agile structural models and internationalized their operations. The UK regards organized crime as a threat to national security which costs the UK at least £32 billion each year, leads to loss of life, and can deprive people of their security and prosperity (https://www.gov.uk/government/publications/official-development-assistance-oda-fco-programme-spend-objectives-2017-to-2018/combatting-serious-and-organised-crime-objectives-2017-to-2018).

With this in mind, consider the policing landscape described in Wexler (2018):

- The nation is experiencing an epidemic of drug abuse fatalities, fueled in part by the internet.
- New types of crime, based on technology, are being invented.
- Local gang members and other criminals have noticed that they can make more money, with less risk of getting caught, and smaller penalties if they do get caught, by using technology.
- Even when technologies are invented to prevent crime, criminals adapt.
- The work of criminal investigators is becoming more complex.
- Police are not getting sufficient help from the technology sector, while criminals are using the "dark web" to make themselves anonymous online.
- Technology is changing the environment every day, and most police agencies are far behind the curve. Because policing in the United States is decentralized, there are thousands of small and medium-size agencies that lack the resources to respond to the changes cited above.
- Crime data systems have not kept pace with changes in crime: A major challenge facing the profession is that we do not have an accurate picture of the problems (Wexler 2018).

Given this threat and risk landscape, it has been argued that the '…science of criminal investigations is changing rapidly, and many law enforcement agencies are not prepared for the changes that are taking place' (Wexler 2018). The inherent complexity associated with serious and organized crime and terrorism calls for intervention and investigative strategies and practices rooted in a science-informed, evidence-based approach. Epidemiology, the science of public health, provides a unique approach that can be leveraged and tailored to support policing operations.

2 Epidemiology

Applying epidemiological principles to law enforcement is not a new concept. As described in Masys (2016a:299), references to terrorism being a "virus" or to al Qaeda "mutating" or "metastasizing" are common, and such terminology is useful when catalyzing the use of an epidemiological framework for investigations.

As described in Brownson and Hoehne (2006), epidemiology is the study of the health of human populations. Its functions are:

1. To discover the agent, host, and environmental factors which affect health, in order to provide the scientific basis for the prevention of disease and injury and the promotion of health.
2. To determine the relative importance of causes of illness, disability, and death, in order to establish priorities for research and action.
3. To identify those sections of the population which have the greatest risk from specific causes of ill health, in order that the indicated action may be directed appropriately.
4. To evaluate the effectiveness of health programs and services in improving the health of the population (Brownson and Hoehne 2006)

Each of these four functions has a direct application toward improving the overall health of the population and reflects an inherent systems view of disease aetiology and intervention strategy design.

As a scientific discipline, epidemiology is replete with sound methods of scientific inquiry at its foundation and supporting causal reasoning and hypothesis testing. Characterized as a data-driven, systematically-applied approach of data collection, analysis and interpretation, epidemiology leverages both quantitative and qualitative methodologies including probability, statistics, grounded theory, content analysis, network analysis and other sound research methods. From a systems perspective, social determinants of health are a key influence on the occurrence of disease and other health-related events searching for the "Why" and "How" of such events which are explored through epidemiological methodologies.

3 Understanding Complexity: Cynefin

In addition to a science-based epidemiological framework, the inherent complexity associated with the policing landscape requires one to broaden the traditional approach to investigation and decision making and form a new perspective based on systems thinking and complexity science (Masys 2016b). The Cynefin framework (Fig. 1) allows one to see things from different perspectives, challenge resident mindsets, assimilate complex concepts, and design intervention strategies for real-world problems and evolving criminal organizations.

As described in Snowden and Boone (2007), the framework sorts the issues into five contexts defined by the nature of the relationship between cause and effect. Four of these—simple, complicated, complex, and chaotic – span the spectrum

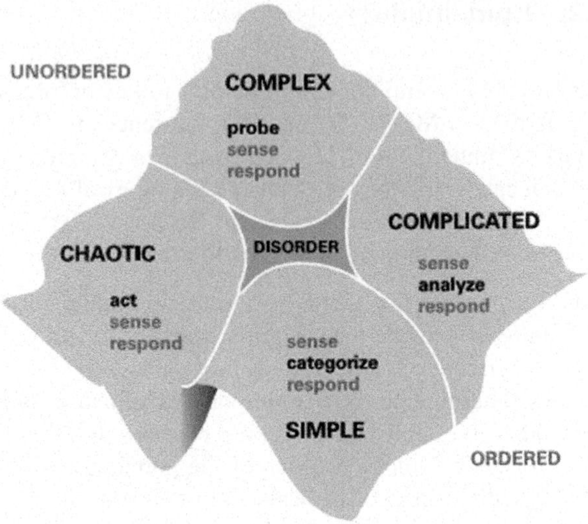

Fig. 1 Cynefin framework. (Snowden and Boone 2007)

from ordered to unordered contexts. The fifth—disorder—applies when it is unclear which of the other four contexts is predominant.

Framing the law enforcement issues along one of the dimensions of the Cynefin implies a specific worldview that shapes how one views the problem space and how one designs intervention strategies. Complex contexts suggest that there is no immediately apparent relationship between cause and effect, and these problems are frequently referred to as wicked or messy because it is difficult to assess the true nature of the problem and therefore how to manage it. Wicked problems (Rittel and Weber 1972) are characterized as nonlinear problem spaces whereby the complex problems are emergent. This is described in Masys (2016b). Given the dynamic and complex policing landscape, problem framing afforded by such approaches as Cynefin is critical for facilitating design of not only investigative tools and approaches, but also development of more strategic intervention strategies to deal with law enforcement issues, and when used in conjunction with an epidemiological framework, can produce manageable investigative methods for use in complex investigations.

4 Connecting Epidemiology and Policing Practices

On the surface, scientific research and law enforcement investigations might appear to have very little in common. Underneath, however, the investigative techniques used in both disciplines bear a striking resemblance to each other. For example, in scientific thinking, a hypothesis is formed, and evidence collected either disproves or supports the hypothesis. Supporting evidence does not necessarily prove the theory correct, but it can add additional credence and plausibility to the hypothesis. In jurisprudence, the concept of "innocent until proven guilty" is very similar. If the theory of a particular criminal's guilt is proposed, evidence that disproves the suspect's guilt is considered definitive, while supporting evidence adds credence to the case for the suspect's culpability without incontrovertibly proving it. In this way, law enforcement investigations have much more in common with scientific investigations than it might seem.

A particularly useful framework for approaching criminal investigations in a scientific manor is epidemiology. Though there are different definitions of epidemiology, at its core, it is the study of patterns of disease. Infectious disease epidemiology, specifically, operates on the premise that disease is not random, that there is a specific cause or set of causes, that it comes from an identifiable source, and that it spreads through identifiable means (Nelson and Williams 2014). These assumptions are not so very dissimilar to investigative assumptions in organized crime or terrorist networks, where crimes do not happen randomly, but instead have a designed purpose/cause, that there is a leader or leaders (source or sources), and that the network spreads outward from this source.

Through the various tools and methods associated with 'applied epidemiology', as described in (Brownson and Hoehne 2006), five core purposes emerge with regards to understanding disease etiology:

1. synthesize the results of etiologic studies to assess cause across a body of literature;
2. describe disease and risk factor patterns to set priorities;
3. develop and evaluate public health programs, laws, and policies;
4. measure the patterns and outcomes of health care; and
5. communicate epidemiologic findings effectively to health professionals and the public.

These epidemiological core purposes map well to the intelligence cycle described in Strang and Masys (2016) illustrated in Fig. 2 (Bruenisholz et al. 2016).

Epidemiologists work across the spectrum from tactical, operational, and strategic policy-related interventions. They can:

- identify potential interventions based on causal criteria.
- design interventions and evaluation protocols.
- assist policy makers in evaluating the effects of the intervention and in formulating broader policies related to the intervention.

In essence, epidemiology is about: surveillance; investigation; analytic studies; evaluation; intervention strategy design and development; and strategic policy development. Supporting these functions are key operational principles that resonate with law enforcement operations:

1. data driven
2. systematic and unbiased approach to collection, analysis, and interpretation of data
3. pattern recognition, cluster analysis

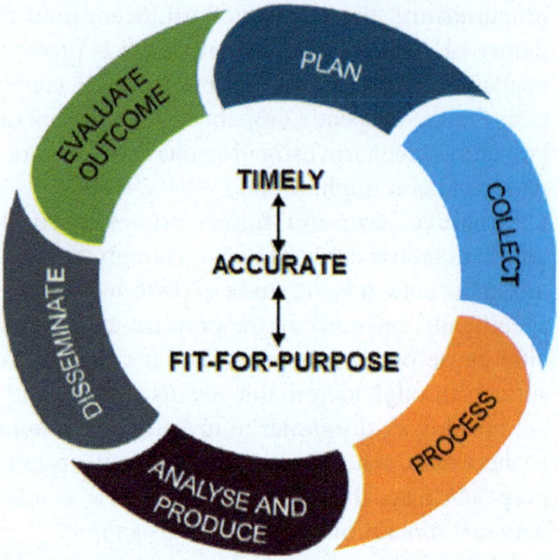

Fig. 2 Intelligence cycle. (Bruenisholz et al. 2016:24)

4. spatial and temporal analysis
5. examining determinants
6. behavior analysis
7. surveillance

5 Applications of Epidemiological Concepts in Policing Operations

In applying epidemiological concepts to policing operations, let's start at the periphery, where most investigations, both epidemiological and criminal, begin. In the criminal world, a crime is reported. In the epidemiological world, a disease is reported. In many instances, this crime and this illness are isolated. The criminal is apprehended or the illness is treated, and in both cases it stops right there. In other instances, the crime that is committed, or the disease that is reported, are routine and very common in that area. Petty theft or the common cold, for instance. These crimes and illnesses are generally taken care of on a routine basis, and they do not raise any undue red flags among investigators.

But sometimes, there can be a sudden rash of burglaries that share a similar modus operandi or a string of drug-related deaths, for instance. Similarly, there can be a sudden rise in illness with similar clinical descriptions or unexpectedly high morbidity. This upswing marks a change from the status quo. It is the first indicator that something new or out-of-ordinary is taking place. In crime, this can indicate a new criminal element moving into an area or new leadership of an old criminal element. In epidemiology, this can indicate the outbreak of a new disease or a genetic shift in an old disease that increases virulence. Either way, the first step is in noticing the shift. Noticing this shift is about having the capability for 'weak signal detection'.

Public health departments have passive surveillance in place that monitors the occurrence of disease within a region. Surveillance in the public health context is not at all like surveillance in the law enforcement context, which involves active observation of subjects, locations, and activities. Public health surveillance instead involves receiving routine reports from doctors, hospitals, laboratories, and public health departments on the occurrence of reportable diseases (Ashbaugh et al. 2017). These reportable diseases have been designated as such because they are illnesses which can potentially become a larger threat to public health, and occurrences of such diseases are entered into a database and evaluated in periodic reports for any observable changes.

Developing a similar database for criminal activity can be more difficult in that, while raw numbers of various crimes are easily analyzed and reported, characteristics of a particular criminal or criminal group can be difficult to distinguish. Disease reports follow a standard case definition and format, and while they can be more or less reliable based on the reliability of the reporting source, criminal reports contain much less standardization of content and much more variability in source reliability. However, a criminal investigator with an index of suspicion can certainly organize and analyze multiple criminal cases with an eye to finding ones with key shared characteristics that would define a particular criminal or criminal group.

Once this change has been noticed, the objective in both types of investigations is to find the source and stop its spread. The source in an outbreak investigation can be an index case or point source. This point source or index case is the one source of introduction of disease into the population, whether it is an individual exposed to a zoonotic virus through contact with the natural animal host, or someone who just landed in a new country after being exposed in another country. In criminal terms, this source can either be a single criminal (i.e. a serial murderer), or someone new arriving in an area. Looked at in terms of a terrorist organization, if a trained operator moves into a new region to begin setting up operations, this operator can be viewed as the point source from which disease will spread. If such a source remains in the environment unchecked, it becomes a continuing common source, spreading disease to multiple people over a prolonged period of time, as would be the case in an established crime syndicate.

Outbreak investigations can also uncover multiple introductions of the same disease or similar strains of a disease into a population congruently or within a relatively short timeframe. Investigating the similarities between these multiple sources can often point back to a common natural reservoir of disease, just as finding the connections between multiple translocated terrorist cells can lead back to a common origin. Additionally, whether it is from one source or multiple sources, once disease begins spreading from person to person without contact with the original source, it is dubbed a propagated source (Nelson and Williams 2014).

Often, the goal of organized crime or a terrorist network, when described in epidemiological terms, would be to become a continuing common source and propagated source and to establish an area of control or dominance. While viruses and bacteria do not display the human motivations of seeking power, dominance, or an end objective, their biological characteristics tend toward maximum replication of themselves at the expense of the host organisms. It is also not suggested that pathogens possess the will to attempt to evade detection or to change their tactics in response to investigators' probes. They do, however, to varying degrees, possess the abilities of antigenic drift (the gradual tendency over time to change outside surface structures so that established immune responses are less likely to recognize them), genetic shift (a sudden, large shift to an entirely new strain), and active efforts to shut down host immune systems so they cannot oppose the infecting pathogen (Nelson and Williams 2014). And so, though it is not consciously driven as it would be in the case of criminals, pathogens display similar behaviors to criminals in order to carry on their activities.

Influenza, for instance, is a virus with a remarkable ability to change two key structures on its surface so that it is not recognized by host immune systems (Heymann 2015). This constant shifting is what necessitates development of a new flu vaccine each year, and when one of these shifts leads to a sudden increase in virulence or mortality, events such as the Influenza Pandemic of 1918 occurs, killing millions worldwide. Similarly, criminals are constantly shifting their tactics and maneuvers to evade detection or established defenses (Phillips 2018). Additionally, criminals can often hide their presence or appear as non-criminals, just as diseases such as typhoid and dengue can often appear without symptoms yet be just as susceptible to spread as symptomatic cases (Heymann 2015). Diseases such as Ebola, however, have not been found to be asymptomatic in humans, and so, just as it is important for outbreak investigators to be extremely familiar with the characteristics of the pathogen they are investigating, it is vital that criminal investigators are familiar with the characteristics of the network they are investigating.

When the source of disease or the leader of a cell are not readily apparent, contact tracing is often required from an identified case back to an index (original) case. Just as tracing the known associates of a lower level criminal who has been detected can lead up the chain to leadership personnel, so contact tracing can backtrack to the original source of infection. This is a labor-intensive process in both criminal and epidemiological investigations, but it is one of the key links in stopping the pathogen/organization. Identifying all of those who have been exposed is vital to determining where the disease has spread and to stopping its further spread (Nelson and Williams 2014).

Rarely is the case of a criminal investigation straightforward from its point of detection to its source and to all of those who have been infected. Often, only some of the pieces of the puzzle can be assembled, but just as in epidemiology, this can potentially be enough to stop the disease. In epidemiology, though it is helpful, it is not always necessary to find the original source in order to stop an outbreak. Cutting the chain of transmission in any part of its cycle can halt the spread of the disease. Similarly, cutting the chain of events in a criminal or terrorist endeavor can also prevent impending crime or attack (Masys 2016c).

Disease transmission requires a viable pathogen, a mode of transmission, and a susceptible host. Disruption of any of these factors halts further infection. Similarly, disruption of any of these factors in a criminal or terrorist network disrupts the intended crime or attack. Terrorist networks and organized crime require money, resources, personnel, and communications in order to function. All of these needs represent vulnerabilities for disruption of the chain of transmission because, without these, the intended crime or attack cannot occur. Such disruption mapping is detailed in Masys (2016c). Figure 3 illustrates the operational views associated with LeT terrorist organization thereby showing leverage points in the terrorist organizational 'business model' for disruption of operations.

Thus, these requirements can be likened to modes of transmission or "vectors". Pathogens require modes of transmission such as mosquitos, direct contact, water, or even air. Without this intermediary, the pathogen cannot get to the host, and infection cannot occur, just as without money, resources, personnel, and/or communication, a crime cannot make it from inception to fruition.

Just as the pathogen in an outbreak requires a mode of transmission to make it to the host, it also requires a susceptible host in order for infection to take place. In epidemiology, hosts can be susceptible for any number of reasons including stress, malnutrition, immunocompromise, and just plain genetics (i.e. genetics cause differences in cellular receptors, and some genetic codes can manifest as immunity to disease because the target cells lack the receptors required to bring a particular virus through their membranes) (Nelson and Williams 2014). Translated into criminality,

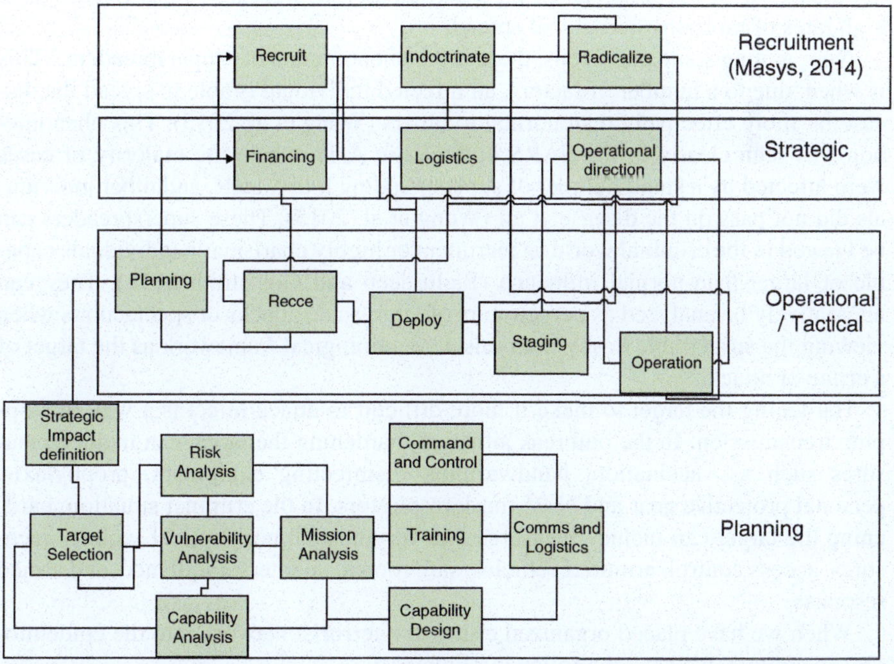

Fig. 3 Operational View (OV-6c) functional operational processes of LeT. (Masys 2016c)

susceptible hosts become targets, and these targets can include victims of the crime/attack or additional recruits to the network.

When looking at recruiting new members, it is useful to look at the characteristics that make these new "hosts" susceptible. Naturally, in areas of endemicity (areas where the pathogen is common or prevalent), there will be more cases of infection, as there are naturally more exposures; just as in areas of high crime or terrorist strongholds, more civilians will become part of the syndicate due to increased exposure to the criminal or terrorist networks within the fabric of the community. But looking at the characteristics of these susceptible recruits often shows commonality in factors such as socioeconomic status, disenfranchisement, and discontent that may lead to vulnerability to radicalization/recruitment (Badurdeen and Goldsmith 2018). Not all recruits come from economically disadvantaged populations, as there are often several highly educated members within terrorist and drug organizations (Bennett, Coleman, and Co. 2012). Identifying the commonalities in these highly-educated or economically advantaged recruits can be just as valuable as identifying the characteristics of the more disadvantaged group. Such indicators regarding counter violent extremism are discussed in detail in (Pressman 2016; Masys 2018).

And just as some pathogens are more virulent than others, some ideologies are more virulent than others, leading to higher and faster rates of transmission. ISIS, for instance, would have very high virulence, with its ideology spreading in pandemic proportions across the globe. And just like the flu, Islamist radicalization is not a new virus, but it has taken on a particularly virulent new form in ISIS with high rates of mortality and global spread.

Now, among some pathogens, there is the phenomenon of "superspreaders." This is when, due to a number of factors, an infected individual is able to spread the disease far more effectively than normal to others (Wong et al. 2015). This phenomenon was seen clearly in the SARS outbreak in Asia, where the majority of cases were infected by a small number of superspreading individuals, and other individuals did not pass on the disease at all (Wong et al. 2015). These superspreaders can be viewed in the criminal world as recruiters or highly charismatic individuals capable of larger-than-normal influence (Badurdeen and Goldsmith 2018). They can alternatively be analyzed as perpetrators of large-scale attacks or spectaculars when viewing the susceptible host within this epidemiological framework as the target of a crime or attack.

Hardening the target to make it more difficult to attack/infect is a way of stopping transmission. In the outbreak situation, hardening the target can include measures such as vaccination, handwashing, disinfecting equipment, prophylaxis, personal protective gear and N-95 mask respirators. In the criminal situation, hardening the target can include measures like varying routine, physical security measures, access control, armored vehicles, safe rooms, countersurveillance, and armed response.

When we have placed organized crime or a terrorist network into the epidemiological context, we have the following linkages:

Outbreak Context	Criminal/Terrorist Context
Outbreak	Organized crime or terrorist network
Detection of disease case	Detection of a crime
Endemicity	High crime area or extremist community
Attack rate/Virulence	Persuasive ideology
Source	Leader
Contact tracing	Known associates
Superspreaders	Recruiters or large-scale attackers
Susceptible host	New recruit or victim/target
Vectors	Money, resources, communications

With these analogies in mind, we can look at analyzing the structures in order to find weaknesses where the cycle can be broken.

Let's take Explosively-Formed Projectiles (EFP's), which became widespread for use against armored vehicle convoys in Iraq beginning in 2005, as an example (Knights 2016). This new form of attack indicated a change from the previously common use of roadside IED's (improvised explosive devices) (Bonsignore and Eshel 2006). Because of their shaped charges, they were much more effective at penetrating the armor on Allied vehicles, and morbidity and mortality from these attacks was much higher. Recognition of the growing use of these devices constitutes an "outbreak," with each of these devices as a case of "disease."

Examination of these devices as they were recovered indicated a uniformity of design, and just as genetic markers on a pathogen would point to a single strain and common source of infection, the conformity of materials and design would point to a single manufacturer. In searching for EFP's that conform to these particular standards, there are, of course, others that can be found which do not meet these exact specifications. These devices would be outside the "case definition" and likely be a product of other bombmakers.

With indications of a single source, investigative efforts would turn toward finding the source in order to halt transmission. This is often a difficult and involved process in epidemiology, and it is even more so in criminal and terrorism investigations. And so, while efforts are ongoing to try to find and remove the source, interim measures can be put in place to stop transmission. As the "host" is most readily available to defending forces, reducing susceptibility of the host can be the easiest cycle interruption to achieve. Measures such as varying travel routes, employing devices to trip passive infrared sensors prematurely, broadcasting signals designed to suppress or prematurely trigger activation frequencies, and changing armor composition are all strategies for disrupting the transmission cycle at the "host" level.

Additional efforts would likely be targeted at modes of transmission and vectors of disease. Discovering how the EFP's are being transported to their destinations and by whom, how they are being financed, where the materials are coming from, and how the orders are being given or decisions are being made for their emplacement are all examples of "vectors" and areas where the chain of transmission can be broken. Temporary removal of any of these vectors forces a change in how the

EFP's are transmitted from production to detonation, and permanent removal of one of the requirements in the mode of transmission would result in the severance of the transmission cycle.

Just as diseases such as Ebola have a suspected but unproven natural source which originates in remote areas that are difficult for researchers to penetrate, so too might the origin of the EFP's be in a region that is difficult for investigators to research (i.e. militant Shia strongholds or Iran). And so, while it may be difficult to eradicate the natural reservoir or source of the EFP infection, location of its introduction into the population is plausible. Focusing on the individuals involved in transmission as "infected cases" and carrying out contact tracing can eventually lead investigators to the original "index cases" within the population (i.e. operatives that are bringing them into the country and distributing them to the Jaysh-al-Mahdi).

Unfortunately, unchecked, this EFP disease, just like terrorist networks and other organized crime, can take root in the new area and become "endemic," where personnel, knowledge, material, and supplies become new sources of EFP production within Iraq (McLeary 2007).

The case of the EFPs exemplifies how applying the epidemiological framework of investigation onto an organized crime element or terrorist network can enable a comprehensive analysis of the criminal elements involved. An epidemiologically-based outbreak investigation would involve the following steps:

- Detect the crime.
- Identify characteristics of the "disease" (case definition).
- Determine if there are similar cases anywhere else.
- Contact tracing – who is involved in or connected to this crime?
- Determine vectors and modes of transmission.
- Identify risk factors (locations or individuals that are likely to become targets of the next crime, or those that are likely to already be involved or become involved in the criminal network).
- Identify the source or sources.
- Plan interventions or treatment.
- Eradicate the source or intervene in transmission.

Many of the steps in this investigative process do no occur in a set order but are rather determined by what information is most readily available. Steps may run concurrently to each other as multiple people work on multiple aspects of the investigation or multiple lines of inquiry are pursued. And as more information becomes available, prior steps are often revisited, with the ultimate long-term goal of finding the source and halting transmission.

An existing analytic structure that is utilized in security and defense circles which lends itself to integration into the epidemiologic framework is the attack cycle (Harding 2014). The attack cycle is illustrative in organizing the phases that an attacker must go through in carrying out an attack (Stratford 2012). These phases include:

1. Preliminary target selection
2. Initial surveillance
3. Final target selection
4. Pre-attack surveillance
5. Planning
6. Rehearsal
7. Execution
8. Escape, explore, exploit

For the purposes of this analysis, these steps can be simplified into:

1. Target selection
2. Surveillance
3. Approach
4. Attack
5. Explore/Exploit

This process can occur very rapidly, in a matter of seconds, or it can be far more involved and require weeks, months, or even years to reach culmination. These steps are most often taught in a defensive framework, where security personnel learn how to interrupt this cycle and prevent an attack. Let's turn this paradigm over, however, and instead use it to outline the phases of a law enforcement attack on an organized crime or terrorist network (i.e. investigation leading to an arrest or disruption of a network).

Target selection would entail the initial detection of a crime or criminal element and the subsequent identification of the unique characteristics of the crime that differentiate it from other endemic crimes. Following this is the surveillance and information-gathering phase in which the bulk of investigatory work is carried out. Tracing known associates, mapping the organization's financial, communications, and resource pathways, analyzing possible targets and spread, and determining the key figures and organizational structure within the element are all investigative methods carried out during this surveillance and information-gathering phase in the epidemiological model.

Once enough information or evidence is gathered, plans for action can be begin. In epidemiology, this is where public health programs and interventions are planned, but in law enforcement, this is where warrants and administrative approvals are requested and approved and plans for effecting arrests are made. These arrests may well not be for the head of the organization but instead for members lower in the hierarchy or involved in one of the vectors of transmission. And so, after the arrest or seizure of evidence has been carried out in the "Attack" phase, "Exploration" is carried out to see where information gained from this attack leads, and the attack cycle immediately resumes with investigative efforts in the "Target Selection" and "Surveillance" phases.

Target Selection	Detect the crime
	Identify characteristics of the crime/disease
Surveillance	Determine if there are similar cases anywhere else
	Contact tracing
	Determine vectors and modes of transmission
	Identify risk factors
	Identify the source or sources
Approach	Plan actions and obtain administrative approvals
Attack	Arrest or remove the source or intervene in transmission
Explore/Exploit	Determine if there are similar cases anywhere else
	Contact tracing
	Determine vectors and modes of transmission
	Identify risk factors
	Identify the source or sources

As shown by the EFP example, the first step when using an epidemiological framework to investigate organized crime or a terrorist network is to detect the crime. On its surface, this may seem like a simple matter, but detection of an organized crime or terrorist-involved incident is not always straightforward. Just as many deadly viruses initially present with flu-like symptoms, organized criminal undertakings can initially blend in with normal crime trends in an area. The Iranian Quds teams, suspects in the above EFP, example often use local crime networks that are already in place, which can initially mask their presence and make detection of the new, more dangerous element difficult (Dai 2018).

Public health systems seek to identify the presence of notifiable diseases as soon as possible through passive, active, and sentinel surveillance systems. Law enforcement has a similar mechanism when it puts out APBs on wanted criminals. These APBs, however, focus on individuals (sources or pathogens) that can be recognized most often by their appearance or distinctive features. Public health systems, however, look for characteristic symptoms that indicate the presence of the disease. In following the epidemiological framework, dissemination of information across law enforcement networks for the type of crime that is likely to be committed by the sought-after criminal may help increase efficacy in detection. Such tactics are used with sensitive substances, like the theft or purchase of large amounts of ammonium nitrates used to make fertilizers or large purchases of controlled prescription narcotics.

Such detection methods can also be used if an investigator has good knowledge of the methods or tactics used by the group he or she is investigating. There are currently limited systems in place to put out wide nets seeking a particular modus operandi, and so current law enforcement must go to increased efforts to share critical information with other law enforcement agencies or jurisdictions in what to be looking for.

There is an overall level of awareness across law enforcement networks about indicators of organized crime or terrorist activity. This can be likened to the passive surveillance that is in place with public health systems, where a healthcare provider reports a case of disease if they come across it. Active surveillance, however, involves the public health investigator actively seeking information, speaking to people, and investigating in an area where a disease presence is suspected. This is what most current law enforcement investigators must do to track down their suspects. Sentinel surveillance, on the other hand, involves priming particular locations and receiving more frequent reports regarding the sought-after disease. In areas where an infectious disease is expected to spread, public health agencies educate and train healthcare professionals in that region on how to recognize early symptoms of the disease and how to report it quickly. Such measures rapidly go into effect at the first hints of viral hemorrhagic fevers in several nations in sub-Saharan Africa, for instance.

Disseminating information across law enforcement networks on the nuances of particular methodologies used by criminal groups, in order for these organizations to be more alert for cases of these crimes, can potentially increase effectiveness in searching for a criminal or criminal group over circulating only pictures and descriptions of suspects. Taking the time to provide education or training on the specifics of what to look for in each case and standardizing how to report and who to report information to can lead to better and faster identification of the sought-after group.

This leads naturally into the surveillance/information-gathering steps of determining if there are any other cases anywhere else and contact tracing. Being able to connect geographically disparate cases allows for investigation of the contacts between these locations and how the translocation has occurred. Such investigation narrows down pools of suspects and leads to additional contacts and vectors. These two steps can also lead to the eventual discovery of the original source or leader of the network, keeping in mind that the etiology of crimes might differ in different locations (just as strains of a pathogen differ across regions and sources) due to their proximity to vectors in the chain of transmission. For instance, contact tracing may lead to discovery of money laundering crimes in conjunction with the financing vector, or it could lead to black market acquisition of illegal explosive materials in conjunction with resource vectors.

The majority of the investigatory activities takes place during the surveillance/information-gathering phase of the cycle, and can potentially lead to predictive analyses in future targets, attacks, recruitment, or activities of the organization (risk factors). These analyses can even have mathematical models, as in the Network- and Agent-Based Models of infectious disease epidemiology. In such models, information about the population, individual, and disease characteristics can be entered into the equation for predictive analysis (Nelson and Williams 2014). This has been applied to the human trafficking problem space in Gardner et al. (2017).

When investigative efforts yield sufficient results, plans for arrests or interventions can begin (the "Approach" phase). In epidemiology, this involves designing education, vaccination, and public health campaigns, but in law enforcement it involves breech plans, obtaining warrants for arrest and collection of evidence, cordon and search plans, or any number of approach tactics. These are then carried out in the "Attack" phase.

While the Attack phase is often the culmination of a tremendous amount of effort, the investigation does not stop there. Instead, the cycle proceeds on to the

Fig. 4 Social ecological model

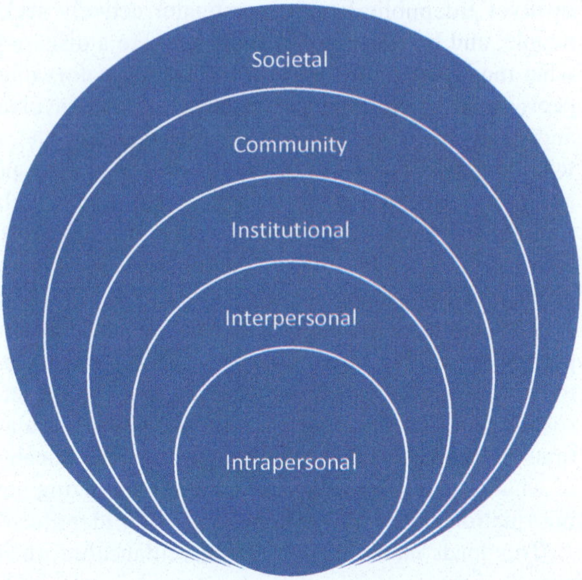

"Explore" phase where information gained from the attack, whether it is through interview of detainees, analysis of seized documents, or examination of forensic evidence, is explored for leads toward further investigative efforts, beginning the Attack Cycle and the epidemiological framework of investigation anew.

Overall, the discussion regarding applications of systems thinking and epidemiology in support of law enforcement operations certainly can be rooted in the social ecological model (SEM) (Fig. 4). As described in (Hayden 2019),

> the basis of the SEM is recognition of the dynamic interplay among the various factors at different levels that affect behavior. Because of this interplay between and among the many levels of influence, changing one can have an impact on them all.

The SEM facilitates understanding regarding disease/criminal aetiology in terms of complexity causality and influence in addition to support in designing intervention strategies. For example, as described in Tsang and Masys (2018), the fentanyl crisis can be framed as a complex problem space that crosses the public health, national security, law enforcement and social services domains thereby aligning intervention strategies within the SEM.

6 Conclusion

As the basic science of public health, epidemiology is the study (scientific, systematic, data-driven) of the distribution (frequency, pattern) and determinants (causes, risk factors) of health-related states and events (not just diseases) in specified populations. Epidemiologists use a systematic approach to assess the What, Who, Where, When, Why, and How of these health states or events. The benefits of adapting and using the epidemiological lens and approach for investigations of organized crime and terrorist networks are described in Stares and Yacoubian (2007:6):

- First, epidemiologists observe rigorous standards of inquiry and analysis to understand the derivation, dynamics, and propagation of a specific disease. In particular, they seek clarity on the origins and geographical and social contours of an outbreak: where the disease is concentrated, how it is transmitted, who is most at risk or "susceptible" to infection, and why some portions of society may be less susceptible, or, for all intents and purposes, immune. Applying the same methodological approach to mapping and understanding [organized crime and terrorist networks] can yield immediately useful guidance on where and how to counter it.
- Second, epidemiologists recognize that diseases neither arise nor spread in a vacuum. They emerge and evolve as a result of a complex, dynamic, interactive process between people, pathogens, and the environment in which they live. Indeed, the epidemiologic concept of "cause" is rarely if ever singular or linear but is more akin to a "web" of direct and indirect factors that play a lesser or greater role in differing circumstances. To make sense of this complexity, epidemiologists typically employ a standard analytical device that "deconstructs" the key constituent elements of a disease. This model helps not only to understand the phenomenon in its entirety but also to anticipate how it might evolve in the future. The same systemic conception of disease can be adapted to understand the constituent elements of [complex police investigations].
- Third, just as epidemiologists view disease as a complex, multifaceted phenomenon, so public health officials have come to recognize that success in controlling and rolling back an epidemic typically results from a carefully orchestrated, systematic, prioritized, multipronged effort to address each of its constituent elements. At the same time, however, it is also recognized that significant progress or major advances can sometimes be precipitated by relatively minor interventions—or "tipping points." Again, there are lessons and insights to be learned here for orchestrating a global counterterrorism campaign (Stares and Yacoubian 2007).

With these reasons in mind, an epidemiological framework can be instrumental in quantifying and conceptualizing the complex situations and relationships involved in organized crime and terrorist network investigations. When implemented successfully, it can ground intangible elements of an investigation, such as "hunches," "intuition," and "gut-feelings" in a legally-defensible, science-based investigatory methodology that assists in systematic exploration of these difficult-to-quantify concepts and connections in both investigation of events which have already transpired, and predictive analysis of future behaviors and outcomes.

References

Ashbaugh, H. R., Kuang, B., Gadoth, A., Alfonso, V. H., Mukadi, P., Doshi, R. H., Hoff, N. A., Sinai, C., Mossoko, M., Kebela, B. I., Muyembe, J. J., Wemakoy, E. O., & Rimoin, A. W. (2017). Detecting Ebola with limited laboratory access in the Democratic Republic of Congo: Evaluation of a clinical passive surveillance reporting system. *Tropical Medicine & International Health, 22*(9), 27. https://doi-org.ezproxy.hsc.usf.edu/10.1111/tmi.12917.

Badurdeen, F. A., & Goldsmith, P. (2018, September). Initiatives and perceptions to counter violent extremism in the coastal region of Kenya. *Journal for Deradicalization*. ISSN: 2363-9849.

Bennett, Coleman & Co. (2012, September 23). *Changing face of terror: The new terrorist is the educated guy from your neighbourhood. Economic* Times. Retrieved from http://link.galegroup.com.ezproxy.lib.usf.edu/apps/doc/A302125910/ITOF?u=tamp44898&sid=ITOF&xid=657d7a5f

Bonsignore, E., & Eshel, D. (2006). Countering the IED threat. (Cover story). *Military Technology, 30*(6), 108–121. Retrieved from http://search.ebscohost.com.ezproxy.lib.usf.edu/login.aspx?direct=true&db=aph&AN=21580349&site=eds-live.

Brownson, R. C., & Hoehne, C. M. (2006). Epidemiology: A foundation of public health. In R. C. Brownson & D. B. Petitti (Eds.), *Applied epidemiology: Theory to practice*. Oxford: Published to Oxford Scholarship Online: September 2009.

Bruenisholz, E., Prakash, S., Ross, A., Morelato, M., O'Malley, T., Raymond, M. A., Ribaux, O., Roux, C. P., & Walsh, S. (2016). The intelligent use of forensic data: An introduction to the principles. *Forensic Science Policy & Management: An International Journal, 7*(1–2), 21–29.

Dai, H. (2018). Tehran's soft power in Africa. *Middle East Quarterly, 25*(4), 1–11. Retrieved from http://search.ebscohost.com.ezproxy.lib.usf.edu/login.aspx?direct=true&db=aph&AN=133938279&site=eds-live.

Gardner, W. L., Panaou, T., Watterson, K., Alao, S., & Jacob, B. G. (2017). Predicting potential human trafficking recruiting geolocations using Eigendecomposed non-zero spatial autocorrelation coefficients and Poissonian heterogeneous gamma distributed variables. *International Journal of Geographic Information System., 4*(2), 1–23.

Harding, D. (2014). Threat management: The coordinated focus on the threat actor, their intentions, and attack cycle. *Journal of Applied Security Research, 9*(4), 478–494. https://doi.org/10.1080/19361610.2014.942825.

Hayden, J. (2019). *Introduction to health behavior theory* (3rd ed.). MA. Jones and Bartlett Learning: Burlington.

Heymann, D. L. (Ed.). (2015). *Control of communicable diseases manual*. Washington, D.C.: Twentieth edition. American Public Health Association.

https://www.gov.uk/government/publications/official-development-assistance-oda-fco-programme-spend-objectives-2017-to-2018/combatting-serious-and-organised-crime-objectives-2017-to-2018

Knights, M. (2016, February 23). *Iranian EFPs in the Gulf: An emerging strategic risk*. The Washington Institute. Retrieved from https://www.washingtoninstitute.org/policy-analysis/view/iranian-efps-in-the-gulf-an-emerging-strategic-risk

Masys, A. J. (2016a). Counter-terrorism and design thinking: Supporting strategic insights and influencing operations. In A. J. Masys (Ed.), *Disaster forensics: Understanding root cause and complex causality*. Cham: Springer.

Masys, A. J. (Ed.). (2016b). *Applications of systems thinking and soft operations research in managing complexity*. Cham: Springer.

Masys, A. J. (2016c). Disrupting terrorist and criminal networks: Crime script analysis through DODAF applications. In A. J. Masys (Ed.), *Exploring the security landscape- non-traditional security challenges*. Springer Publishing.

Masys, A. J. (2018). Radicalization and Recruitment – A systems approach to understanding violent extremism: New Developments through FRAM. In F. Stowell (Ed.), *Systems Research for Real-World Challenges*. Hershey: IGI- Global.

McLeary, P. (2007, April 10). *Get the facts straight on Iran and EFPs*. Columbia Journalism Review. Retrieved from https://archives.cjr.org/behind_the_news/get_the_facts_straight_on_iran.php

Nelson, K. E., & Williams, C. M. (2014). *Infectious disease epidemiology*. Burlington: Jones & Barlett Learning.

Phillips, B. J. (2018). Terrorist tactics by criminal organizations: The Mexican case in context. *Perspectives on Terrorism, 12*(1), 46. Retrieved from http://search.ebscohost.com.ezproxy.lib.usf.edu/login.aspx?direct=true&db=edsjsr&AN=edsjsr.26343745&site=eds-live.

Pressman, D. E. (2016). The complex dynamic causality of violent extremism: Applications of the VERA-2 risk assessment method to CVE initiatives. In A. J. Masys (Ed.), *Disaster forensics understanding root cause and complex causality*. Cham: Springer.

Rittel, H. W. J., & Weber, M. M. (1972). Dilemmas in a general theory of planning. *Policy Sciences, 4*(2), 155–182.

Snowden, D. J., & Boone, M. E. (2007). A Leader's framework for decision making. *Harvard Business Review, 85*(11), 68–76.

Stares, P. B., Yacoubian, M. (2007) *Terrorism as a disease: An epidemiological model for countering Islamist extremism*. https://www.files.ethz.ch/isn/46161/Stares_Terr_Disease_2007.pdf

Strang, S., & Masys, A. J. (2016). Supporting intelligence analysis through visual thinking. In A. J. Masys (Ed.), *Applications of systems thinking and soft operations research in managing complexity*. Cham: Springer.

Stratford. (2012, February 23). *Defining the terrorist attack cycle*. Retrieved from https://worldview.stratfor.com/article/defining-terrorist-attack-cycle

Tsang, C., & Masys, A. J. (2018). Fentanyl crisis: A national SECURITY matter. In A. J. Masys (Ed.), *Security by design*. Springer Publishing.

Wexler, C. (2018). *Critical issues in policing series new national commitment required: The changing nature of crime and criminal investigations police executive research forum*, Washington, D.C. 20036 retrieved from https://www.policeforum.org/assets/ChangingNatureofCrime.pdf

Wong, G., Liu, W., Liu, Y., Zhou, G., Bi, Y., & Gao, G. F. (2015). MERS, SARS, and Ebola: The role of super-spreaders in infectious disease. *Cell Host & Microbe, 18*(4), 398–401. https://doi.org/10.1016/j.chom.2015.09.013.

Sex Offenders' Forensic Awareness Strategies to Avoid Police Detection

Julien Chopin and Eric Beauregard

Abstract In criminology, the rational choice perspective assumes that offenders decide whether or not to commit a crime by weighing the effort, rewards, and costs involved in alternative courses of action. One of the most important costs for offenders is to be identified and arrested by police. In order to avoid arrest, offenders have to rely on specific strategies. In sexual assaults, in addition to appropriate target selection, offenders must make sure to avoid police detection by protecting their identity and cleaning up or destroy forensic evidence that may be left at the crime scene and that can directly lead to their identification. This behavior is known as 'forensic awareness'. Research on forensic awareness strategies during the crime-commission process is scant. A number of studies looking specifically at sexual crimes identified certain strategies related to forensic awareness. According to the rational choice perspective, the use of strategies to avoid police detection should have an impact on the crime solving by police. Few studies have investigated this issue and findings are quite unexpected. The aim of this chapter is to review some of the most important empirical knowledge on the use of forensic awareness strategies in sexual crimes.

Keywords Forensic strategies · Sexual crime · Offender behavior · Police strategies · Criminal investigation

1 Introduction

In criminology, the rational choice perspective (Cornish and Clarke 1986, 1987) assumes that an offender decides whether or not to commit a crime by weighing the effort, rewards, and costs involved in an alternative course of action. One of the most significant costs for offenders is to be identified and arrested by the police. Thus, offenders will make rational decisions to increase the benefits of their crimes while avoiding police detection. According to Cornish (1994), the crime-commission process involves different stages: the pre-crime, crime and post-crime phases. The

J. Chopin (✉) · E. Beauregard
School of Criminology, Simon Fraser University, Burnaby, BC, Canada
e-mail: julien_chopin@sfu.ca; ebeaureg@sfu.ca

B. Fox et al. (eds.), *Science Informed Policing*, Advanced Sciences and Technologies for Security Applications,
https://doi.org/10.1007/978-3-030-41287-6_3

post-crime phase deals specifically with how offenders 'exit' the crime scene and attempt to avoid detection (Cornish 1994). In order to avoid arrest, offenders have to rely on specific strategies. However, there are potentially as many strategies as there are different types of crimes and criminals. In sexual assaults, in addition to appropriate target selection, offenders must make sure to avoid police detection by protecting their identity, and cleaning up or destroying forensic evidence that may have been left at the crime scene that can directly lead to their identification. In those cases, offenders will show more or less 'forensic awareness'. With the development of technologies and forensic sciences, the use of forensic evidence in criminal investigations has increased significantly (Beaver 2010; Strom and Hickman 2010). Therefore, offenders adapt their modus operandi to hide evidence in order to ultimately avoid apprehension. This behavior is known as 'forensic awareness' (Davies 1992).

In this chapter, we discuss the theoretical foundations of rational choice perspective to explain offenders' behavior to avoid police detection. Second, we discuss the two major hypotheses explaining what influence the use of forensic awareness strategies (FAS) by sex offenders, namely the CSI effect and the learning perspective through previous contact with the criminal justice system. Third, this chapter discusses why some sex offenders do not use FAS to avoid police detection. Fourth, we provide a description of the type of FAS used by sex offenders and discuss various empirical findings. Fifth, we review the few studies analyzing the relationship between the use of FAS and crime/offender characteristics in a criminal profiling perspective. Sixth, we discuss the influence of the FAS on the crime solving for rape cases as well as for sexual homicides. The final section deals with the criminal investigation, and presents findings on how investigators manage forensic evidence and how they can have an impact on the solving of a crime.

2 Avoiding Police Detection for Sex Offenders: A Rational Choice Perspective

In criminology, the rational choice approach (Cornish and Clarke 1986, 1987) is a theoretical perspective used to explain at a macro level, the crime commission, and at a micro level, the decision-making process of individual during the crime. At the macro level, the rational choice perspective assumes that offenders decide whether or not to commit a crime by weighing the effort, rewards, and costs involved in alternative courses of action. At the micro level, this theoretical framework assumes that each decision taken by offenders during the crime-commission follows a cost-benefit analysis to increase the crime success and decrease the risks associate with the crime-commission. Rational choice theory does not assume that offenders are rational individuals. This framework only assume that offenders follow a logic from their point of view to increase the benefit and decrease the risks. Several constraints must be taken into account to analyze an offender's decision process under the lens of rational choice perspective. Rational decisions taken by offenders can be considered as subjective because they depend, on the one hand, on internal constraints (i.e. individual characteristics) and, on the other hand, on external constraints (e.g., situational factors, crime characteristics, etc.). Internal constraints concern individuals' pre-existing cognitive schemas about themselves, others, and the world (see Polaschek and Ward 2002; Ward 2000). External constraints concern crime process hazards and the fact that only partial information is available for offenders. Several studies applied the rational choice approach to sexual crimes and found that sex offenders follow a rational choice reasoning to commit their crimes and achieve their objectives (see for example Beauregard 2005; Beauregard and Leclerc 2007; Beauregard et al. 2007a, b, 2010; Chopin 2017; Chopin and Beauregard 2019a; Chopin et al. 2019; Chopin and Caneppele 2019a, b; Murray 2007). As mentioned previously, the post-crime stage deals specifically with how offenders 'exit' crime scenes and attempt to avoid detection (Cornish 1994), while FAS can be used during each stage by offenders.

The concept of crime achievement can be related to the rational choice perspective and is based on the idea that each crime pays, and focus especially on the payoffs and its beneficiaries. In criminology, it has been mainly applied to crimes with economic outcomes (Bouchard and Nguyen 2010; Morselli and Royer 2008; Morselli and Tremblay 2004; Morselli et al. 2006; Nguyen and Bouchard 2013; Tremblay and Morselli 2000). Crime achievement is evaluated in terms of productivity, performance, and efficiency to secure the payoffs purchased by offenders and was applied to the field of sexual crimes (Lussier et al. 2011). For sexual offenses, Lussier et al. (2011) gave a definition of what constitutes achievement for sex offenders. Two different types of achievement are mentioned: 'the ability to maximize payoffs from sexual crimes, and the ability to avoid police detection and sanctions' (Lussier et al. 2011, p. 435). Avoiding police detection is part of a successful sexual crime (Lussier et al. 2011; Lussier and Mathesius 2012) and constitutes a major part of the equation to understand the crime-commission process. Despite this general theoretical framework, there are variations in the use of FAS by sex offenders – some of them are especially forensically aware- while other are not trying to avoid police detection. To explain this variance between individuals, two general hypotheses have been formulated.

3 What Lead Sex Offenders to Avoid Police Detection?

Several studies highlighted that since the end of the twentieth century, the use of forensic evidence in criminal investigations has significantly increased (Beaver 2010; Strom and Hickman 2010). To understand what influences sex offenders to use FAS, two general hypotheses have been highlighted in the research. The first one has been labeled as the 'CSI effect', while the second suggests that FAS is a feature of experienced sexual offenders.

3.1 CSI Effect

The CSI effect hypothesis suggests that the amount of publicity given to the importance of evidence used to apprehend criminals maybe increasing offenders' recognition of the need to become forensically aware (Cole 2010; Cole and Dioso-Villa 2006). This amount of publicity would be due to the increase of crime drama TV shows since the beginning of the 2000s and has resulted in the increase use of FAS and the sophistication of offenders to avoid police detection. This hypothesis has not been empirically supported for sexual crimes.

Findings of various studies have shown that the use of FAS is not a constant for sex offenders, and that generally they use unsophisticated strategies (Beauregard and Bouchard 2010; Beauregard and Martineau 2014; Brown and Keppel 2012; Park et al. 2008). In their study, Beauregard and Bouchard (2010) found that only a minority of rapists used FAS to avoid police detection. They also found that most rapists used unsophisticated strategies. Very few offenders took precautions involving the use of forensic evidence that could contain traces of DNA from the crime scene. These findings have been confirmed by Park et al. (2008) for serial rapists and by Beauregard and colleagues for sexual homicide (SH) (Beauregard and Martineau 2013, 2014; Chopin and Beauregard 2018, 2019b, c). Chopin and Beauregard (2018), in their study comparing cases of SHs with violent rapes and non-sexual homicides, found that violent rapists and sexual homicide offenders (SHOs) used FAS similarly but are less likely to avoid police detection by destroying forensic evidence compared to non-sexual murderers.

As suggested by Beauregard and Martineau (2017), most of the TV shows that could have increased the use of FAS by offenders appeared after 2000. A time series analysis of the average number of FAS used by Canadian SHOs over the last half of a century showed that the use of detection avoidance strategies began to increase significantly during the 1970s. They suggested that the FAS use evolution cannot confirm the CSI hypothesis. Figure 1 provides a temporal analysis of the average number of FAS used by French stranger rapists over the last half of a century, while Fig. 2 provides similar analysis for French SHOs. The use of FAS used by rapists follows a very stable trend suggesting that no CSI effect occurred for stranger rapists. For SHOs, the trend follows findings of Canadian offenders with a first increase in the 1990s and a second increase at the beginning of the 2000s. Although the second increase could have been due to the CSI effect, we notice that the use of FAS decreases drastically until 2010. However, the number of crime drama TV shows did not decrease during this time, which goes against the CSI hypothesis. Following the conclusion from Beauregard and Martineau (2017, 2018), no evidence to date has allowed us to confirm the CSI hypothesis.

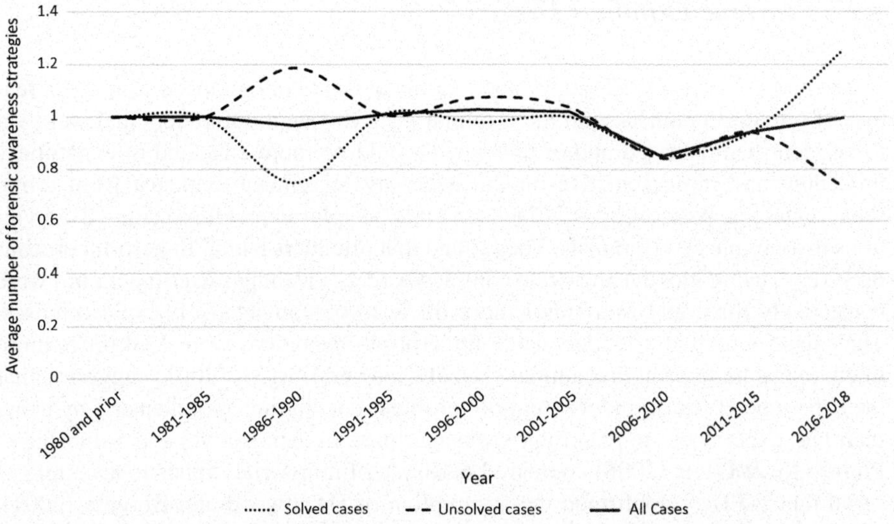

Fig. 1 Average number of forensic awareness strategies used by stranger rapists from 1979 to 2018 in France for solved (n = 3419) and unsolved (n = 1191) cases

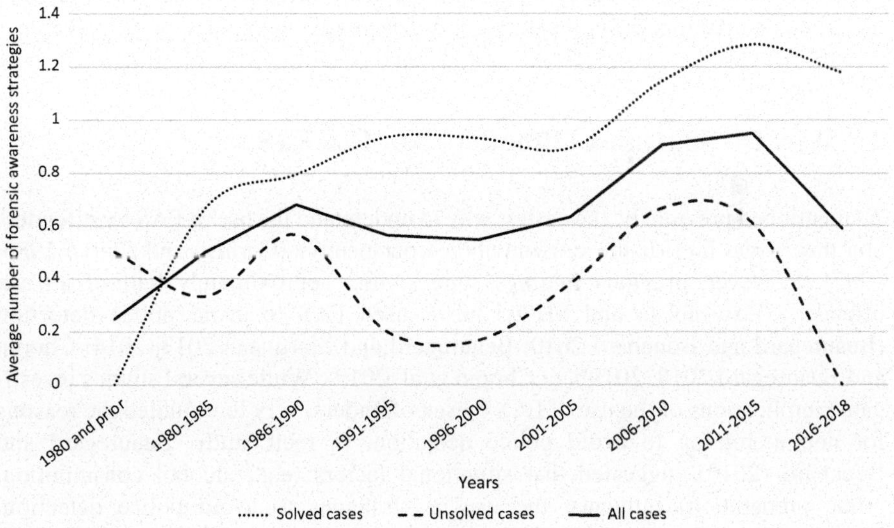

Fig. 2 Average number of forensic awareness strategies used by sexual murderers from 1979 to 2018 in France for solved (n = 412) and unsolved (n = 81) cases

3.2 *Previous Criminal History*

In her study, Davies (1992) suggested that the use of FAS by rapists could be a feature of previous contacts with the criminal justice system. Offenders that were yet to be identified, arrested and/or convicted would be more educated in evidentiary procedure and more sensitive on the necessity to remove evidence from crime scenes. Davies, Wittebrood, and Jackson (1997) explored previous criminal convictions in a sample of 210 rapists. They found that offenders taking fingerprint precautions (e.g., using gloves) and destroying DNA (e.g., cleaning the crime scene) were respectively three and four times more likely to be convicted of sexual offenses. They also found that offenders using fingerprint precautions were 13.4 times more likely to use fingerprint precautions. Hazelwood and Burgess (2001) suggested that the experienced sex offenders are more proficient in eluding apprehension by using their past experiences to plan future sexual crimes. In their study, Park, Schlesinger, Pinizzotto, & Davis (2008) compared a sample of non-serial rapists to a sample of serial rapists. They confirmed the assumption of Hazelwood and Burgess (2001), and showed that serial rapists are more likely to use FAS compared to non-serial rapists.

If previous criminal convictions seem to have a certain impact on the forensic awareness sophistication, a significant number of sex offenders do not use FAS after the crime completion. Several explanations have been highlighted in the research.

4 Why Do Some Sex Offenders Not Use FAS?

As mentioned previously, the easiest way to understand the use of FAS by offenders at a theoretical level is the rational choice perspective (Cornish and Clarke 1986, 1987). However, previous findings suggest that approximately half of all sex offenders (i.e., rapists and SHOs) never used FAS to avoid police detection (Beauregard and Bouchard 2010; Beauregard and Martineau 2013, 2014; Chopin and Beauregard 2018, 2019b, c; Chopin et al. 2019). While several studies investigated implications of the use of FAS by sex offenders, very few studied the reasons for not attempting to avoid police detection. In their study, Beauregard and Bouchard (2010) suggested that situational factors (e.g., alcohol consumption, crime preparation) influence the use of strategies to avoid police detection. Beauregard and Bouchard (2010) discussed the fact that the use of alcohol and/or drug by sex offenders is relatively common prior to sexually assaulting someone (see for example Brecklin and Ullman 2010; Chopin 2017; Chopin and Beauregard 2018; Ciavaldini 1999; Testa and Livingston 2000; Zawacki et al. 2003) and could affect their decision-making, as an intoxicated offender will likely overlook some important details to avoid police detection. In their study on burglars, Cromwell and Olson (2004) observed a similar pattern that they called classic inverted U-shaped dose response performance curve. This phenomenon suggests that offenders are aware until the crime was effectively committed. After they obtained their gratification, they focus on leaving the crime scene and are less likely to use FAS to avoid police detection. In their study, Beauregard and Bouchard (2010) interviewed rapists about the reasons why they did not use FAS. The rapists mentioned the following explanations:

- The crime was not planned, it was spontaneous, and they did not think about FAS;
- It was their first offense and they had no idea about what to expect in regards to the potential risks to be identified;
- Mental status was altered during crime due to alcohol, drugs, or extreme emotions (e.g., anger). They could not think about the use of FAS due to their state of confusion;
- They thought that the victim was willing and they did not need to avoid police detection;
- Some part of the crime process was incompatible with the use of FAS (e.g., for vengeance, it was important that victims saw their face);
- An alternative to FAS was used.

These findings show that despite their use in the crime-commission process, especially for avoiding police detection, some offenders never use strategies to avoid police detection, while others do. Among the group of sex offenders who are forensically aware, there is a myriad of practices and techniques that they use.

5 How Do Sex Offenders Use Forensic Awareness?

Several studies focusing on the type of FAS used by sex offenders tried to identify, describe, and classify them. Generally, research made a distinction between strategies to protect the offender's identity and strategies to destroy forensic evidence. In one of the first studies discussing FAS of sexual offenders, Davies (1992) described what she operationalized as the forensic awareness concept. She described firstly offender behaviors to avoid interruption. Among these behaviors, Davies (1992) described situations where offenders gagged their victims, targeted women who were alone, and disconnecting their phone. In cases occurring outdoor, victims were usually forced to go to a place with a lesser risk of interruption. Second, Davies (1992) described offenders' behaviors to protect their identity. Included in this category are behaviors made to prevent the victim from seeing the offender's face (e.g., use of mask, attack after nightfall, using verbal instruction about where victim should look, etc.). Sex offenders also try to protect their identity by avoiding leaving behind biological evidence (DNA, fingerprints) at the crime scene (e.g., use of condoms, wearing gloves, etc.) and lying to mislead crime investigators (e.g., giving a false name). Third, Davies (1992) identified that sex offenders, particularly the more experienced, threaten their victims to not report the crime by using their identifying personal information to locate them later (e.g., looking for address and name on pieces of identification). Finally, the last types of strategies used by sex offenders occur at the end of crime. The main objective is to ensure a safe departure from the crime scene (e.g., threatening the victim not to move and not to look where the offender leaves). In their study based on a sample of 210 stranger rapists, Davies et al. (1997) found that 15% of offenders wore gloves, 5% removed semen from the crime scene, 28% hid their face, and 20% lied about their name. In the study by Park et al. (2008) based on a sample of 44 serial rapes committed by 22 serial rapists and 22 single-victim rapists, findings show that serial rapists are more forensically aware (39%) than non-serial ones (13.6%). However, there was no significant differences between the two samples for removing semen from the crime scene. In their study based on a sample of 222 rapes, Beauregard and Bouchard (2010) found that 55.9% of rapists in these cases concealed their identity, 44.1% wore gloves, 32.9% prevented their faces from being seen, 6.8% wiped semen and only 1.8% wore a condom, while 1.4% lied about their name. In the study by Chopin et al. (2019) based on a sample of 4354 rape cases (i.e., 3243 solved cases and 1111 unsolved cases), they found that the most prevalent FAS used by rapists are use of threats, covering the victims' mouth, the use of condoms, masks, and lying about their identity.

As to sexual homicide (SH), some research has tested the use of FAS by offenders. In the studies by Beauregard and colleagues (Balemba et al. 2014; Beauregard and Martineau 2013, 2014, 2016) based on a sample of 350 SHs, findings indicated that 54.3% of SH offenders (SHOs) never used FAS to avoid police detection. They also found that 30.6% of Canadian SHOs destroyed or removed evidence, 11.4% acted on the victim, and 4.3% protected their identity. They also observed that 24.9% of SHOs used only one FAS, 11.7% used two strategies and 9.1% used 3 or more FAS. In the study by Chopin and Beauregard (2019b) comparing French (n = 412) and Canadian (n = 350) cases of SHs, findings showed that 57.0% of French SHOs used at least one FAS. They destroyed or removed forensic evidence in 30.6% of cases, acted on the victim in 29.1% of cases, protected their identity in 14.1% of and staged the crime scene in 7.0% of the cases. Bivariate analysis showed that French SHOs protected their identity and acted on victims more often than Canadian SHOs, while they less often disposed of the victim's body. In their typology of SHs of children, Chopin and Beauregard (2019c) found that the intentional/prepubescent and inadvertent/preteen SHO categories were significantly more forensically aware than others. Approximately 80% of the Intentional/prepubescent SHOs used at least one FAS while 100% of the inadvertent/preteen SHOs did. However, in 50% of the cases semen was found by the police for crime committed by inadvertent/preteen SHOs.

6 Is the Use of FAS a Specific Feature of Sexual Offenders?

Only a few studies analyzed the relationship between crime/offender characteristics with the use of FAS. However, these findings can improve suspect identification by the police. In the study by Davies and Dale (1995) based on a sample 79 solved serial and single-victim rapes, their findings suggested that rapists traveling longer distances were more likely to avoid police detection. Beauregard and Field (2008) analyzed body disposal patterns from a sample of 85 SHs and found that forensically aware offenders were more likely to move the victim's body than others. In a different study, Beauregard and Martineau (2016) tested the assumption that organized SHOs are better at avoiding police detection than disorganized SHOs. They analyzed a sample of 350 Canadian SHOs using variables associated with the organized type described by the FBI (e.g., target the victim, used a con approach, moved the victim body, used restraints, took items from victims, etc.). Results revealed that some of the organized behaviors were associated with avoiding (i.e., offender moved the victim's body, offender selected less risky locations) or delaying police detection (i.e., offender had vaginal intercourse with victim, evidence of overkill, offender concealed the victim's body, offender selected less risky locations), but the majority of organized behaviors had a counterproductive effect and reduced the delay of police detection (i.e., offender targeted the victim, offender used restraints, offender had anal intercourse with victim, offender used a weapon, offender moved the victim's body). With the same sample, Reale, Beauregard and Martineau (2017) tested if sadistic SHOs were more forensically aware. Their findings showed that investigative awareness was a characteristic of sadistic SHOs. They found that sadistic offenders used strategies to avoid police detection more often than non-sadistic SHOs. Specifically, they found that sadistic SHOs selected deserted places to encounter their victims and dump their bodies, in addition to using FAS. They acted on victims and/or the environment (e.g., disabling the victim's phone or car, or using police scanners to track police movement), destroyed and removed forensic evidence (e.g., cleaning or setting fire to the crime scene), and used other precautions such as staging the crime scene or protected their identity (Reale et al. 2017).

7 Does the Use of FAS Allow Sex Offenders to Avoid Police Detection?

A few studies analyzed the implications of FAS on crime solving. To the best of our knowledge, only three studies investigated this question and focused on SH cases (Balemba et al. 2014; Beauregard and Martineau 2014; Chopin et al. 2019). Beauregard and Martineau (2014) analyzed a sample of 350 SHs committed between 1948 and 2010 in Canada. They compared 250 solved cases (i.e., the offender has been identified) with 100 unsolved cases. In their study, they tested the influence of a set of variables, including forensic awareness strategies, on the status of crime (i.e., solved, unsolved) and the number of days before body recovery. Their results were quite surprising because they suggested that cases of SH involving the use of at least one FAS to avoid police detection were more likely to be solved by the police (Beauregard and Martineau 2014). They also found that SHOs who did not leave biological evidence at the crime scene were more likely to be identified by police (Beauregard and Martineau 2014). Beauregard and Martineau (2014) suggested that these results could be due to the specific nature of police data concerning SHs. The lack of victim and witness testimony limited the data related to FAS used by offenders, especially for unsolved cases. It is also possible that FAS used by SHOs are difficult to identify and that the ones observed by the police are not the most effective to avoid police detection (Beauregard and Martineau 2014). In their study, Chopin et al. (2019) analyzed the influence of victimology factors, modus operandi characteristics, and FAS on the rape case status (solved, unsolved). They used a sample of 4354 cases of rapes committed by offenders that were unknown to the victims at the time of the offense (3243 solved cases and 1111 unsolved cases). Their findings showed that the use of FAS by rapists had a very weak influence on the case status compared to victimology characteristics (i.e. crime conditions linked with victims' lifestyle) and modus operandi characteristics (especially pre-crime phase). Contrary to Beauregard and Martineau (2014) findings on SHs, Chopin et al. (2019) found that some FAS used by rapists are in fact efficient to avoid police detection. They found that rapists using strategies to protect their identity were more likely to avoid police detection. Among this cluster of strategies, wearing a mask during the rape was a better predictor of avoiding police detection than to worry about leaving forensic traces at the crime scene (Chopin et al. 2019). When the rapist is completely unknown, the investigators will need to rely mainly on the victim's description of her assailant, which will be limited in cases where offenders wear a mask. The fact that some sex offenders leave biological evidence at the crime scene is not sufficient to solve rape cases. When a DNA sample is collected it must be compared with other samples from a DNA database. If the required information is not available in the DNA database, there is nothing to compare it to and the biological traces cannot necessarily contribute to solving rape cases (Chopin et al. 2019).

Similar to SH cases (Beauregard and Martineau 2014), the study by Chopin et al. (2019) found that some FAS used by rapists lead to an increase in the risks of being arrested. Rapists that lie about their name or threaten the victim were more likely to be identified. This suggests that the verbal exchange between offenders and victims allows victims to collect information about the offender that could lead to his arrest (Chopin et al. 2019).

In their study, Balemba et al. (2014) used the same sample of SH cases as Beauregard and Martineau (2014) (i.e., 250 solved cases and 100 unsolved cases) to establish two classifications examining detection avoidance strategies. The first classification was based on the whole sample (solved and unsolved cases) and was divided in three groups: Sloppy/Reckless, Violent/Sadistic and Forensically Aware. The first two clusters (i.e., Sloppy/Reckless and Violent/Sadistic) involved offenders that are often detected by the police (respectively 83.3% and 80.5%). For the last cluster including forensically aware SHOs, half of the cases were solved by the police. The Sloppy/Reckless offenders were characterized by committing multiple sexual acts on victims and usually leaving semen at the scene, suggesting they do not use condoms or do not use it correctly (Balemba et al. 2014). The Violent/Sadistic offenders were more focused on violence and torture than sexual acts. They were more likely to be detected through stealing victim trophies, using a high level of violence, and leaving biological evidence on or near the victim (Balemba et al. 2014). The last cluster of the general sample (i.e., solved and unsolved cases) included SHOs that were the most difficult to detect (half of the cases remained unsolved). These offenders did not have vaginal and/or anal intercourse and never left semen at the crime scene. These offenders executed their crimes in a way that decreased the risk to be detected.

Although previous studies suggest that some FAS may help the offender to avoid police detection, it is important to consider what implications arise for the police during criminal investigations.

8 Criminal Investigations and Forensic Identification

When a crime is committed, forensic science theory suggests that physical evidence is left at the crime scene and can be exploited by the investigators. In practice, evidence detection by police investigators is not as easy as suggested by theory, and a limited number of forensic materials are actually collected at crime scenes. Research has also highlighted an attrition process from collected forensic samples at crime scenes to the actual analysis in laboratories. Finally, investigative failures can lead police to fail in solving crimes.

8.1 Theory of Forensic Evidence in Sexual Crime Investigation

In their study, Johnson et al. (2012) described a general theory of forensic evidence based on Locard's exchange principle (Locard 1920). This basis of the forensic science suggests that during physical activities involving two (or more) individuals (i.e., offender and victim), a reciprocal transfer of physical material between them occurs (Johnson et al. 2012). In that perspective, the objectives of forensic identification are to detect and examine the physical material found at crime scenes and to provide information to investigators and the courts to identify, arrest, and convict offenders. Specifically, this information helps to describe the crime-commission process, to establish the identity of the offenders and the victims (in case of SHs) and to corroborate criminal investigation information.

Peterson et al. (1984) described the reasons why forensic evidence was used: To establish an element of the crime, to identify a suspect or victim, to establish an association, to test statements and alibis, to reconstruct the scene, and to corroborate the facts of the case. In cases of sexual crime, Locard's exchange principle (Locard 1920) is particularly important because it involves physical violence (e.g., victim resistance, blow given by offenders, weapons, etc.), material (e.g., restrains, foreign objects used for sexual purpose, etc.) and body fluid (e.g., semen, blood). Beauregard and Martineau (2018) described the role of forensic evidence in case of rapes specifically:

1. To search for the presence of semen from inside a rape victim to prove penetration.
2. To link the offender to the weapons used during crime.
3. To analyze a fingerprint, which may be used to identify the offender or contradict the offender's statement.
4. The use of blood splatter analysis to indicate where the offender and victim were standing at the time of impact (likely by a blunt object, knife, or gun shot).
5. To identify fingerprints and semen at a rape crime scene and analyze it in order to associate the suspect with the location and corroborate the victim's statement.

8.2 Detection of Forensic Evidence in Practice

In their study, Peterson et al. (1984) suggested that in cases of sexual crimes, the most common evidence collected at crime scenes are the fingerprints, bodily fluids, and hair. In an important study, Johnson et al. (2012) analyzed a sample of 602 cases of rape that occurred in 2003. Their findings (see Table 1 below) showed that forensic evidence was collected in 63.8% of cases. Biological evidence (e.g., blood, saliva, DNA, condom, semen, vaginal, urine, bite mark, etc.) was collected in 53.5% of cases, latent prints (i.e., fingerprints, palm prints) were collected in 4.3% of the cases, pattern evidence (i.e., footprint, footwear) were collected in 4.3% of the cases, traces (e.g., hair, pubic hair, plastic, glass, etc.) were collected in 4.2% of the cases, and natural/synthetic materials (i.e., bindings, clothing, bed/bath, carpet) were collected in 42.2% of the cases.

To help investigators collecting evidence after a rape, sexual assault kits (SAK) can be used. SAKs guide and standardize the collection of forensic evidence (biological and trace evidence) by medical personnel to corroborate sexual contact (Du Mont and Parnis 2000; Johnson et al. 2012). In their study, Johnson et al. (2012) found that SAKs were used in 51.3% of the cases but were actually analyzed in 5.3% of the cases. McGregor et al. (2002) analyzed the efficiency of SAK and found that sperm or semen was identified in 38.2% of the cases where a SAK was used. In terms of implications for arrest and charges, they mentioned that the forensic exam is important for criminal investigations but played a limited role to actually charge offenders (McGregor et al. 2002).

8.3 Police Efficiency to Detect and Analyze Forensic Evidences

Johnson et al. (2012) mentioned that several factors can affect the success of the forensic investigation. Among these factors is the type, quantity, and quality of evidence, and the laboratory level of expertise (Fisher 2003; James and Nordby 2005; Johnson et al. 2012). Samples taken and analyzed are compared with existing information contained in databases and can lead to an association or exclusion of a suspect. Researchers highlighted the discrepancy between the collection and the

Table 1 Analysis of collected, submitted and examined evidences in rape cases

	% of collected evidences	% of submitted evidences in comparison with collected evidences	% of examined evidences in comparison with collected evidences
Total	63.79%	50.52%	29.17%
Biological	53.49%	42.24%	27.64%
Finger prints	4.32%	96.15%	69.23%
Pattern evidence	0.66%	0.00%	0.00%
Firearms/ weapons	1.83%	9.09%	9.09%
Natural/ synthetic materials	42.19%	28.74%	7.87%
Generic objects	2.49%	0.00%	0.00%
Electronic/ printed data	1.50%	0.00%	0.00%
Trace	4.15%	72.00%	24.00%

Adapted from Johnson et al. (2012)

analysis steps. Johnson et al. (2012) analyzed the follow-up of forensic evidence collection and the analysis process. Their main findings are summarized in Table 1. We observe that generally, evidence was collected in more than half of rape cases (63.8%) while among this collected evidence, only 50% were submitted to a laboratory and 29.2% were analyzed. The most often collected evidence are biological, and natural/synthetic materials (respectively 53.5% and 42.2%). Moreover, the most often examined evidence in laboratories are fingerprints, while they are among the less collected evidence.

The collection of forensic evidence – or lack thereof – may impact crime solving. Moreover, it has been shown that sometimes this whole process may lead to investigative failures.

8.4 *Investigative Failures*

In their study using a sample of 350 cases of solved and unsolved SH cases, Balemba et al. (2014) found that a segment of the unidentified offenders were highly forensically aware, whereas some other offenders (approximately one third) were simply lucky in not being caught. As suggested by Rossmo (2009, p. 3), despite the fact that crime characteristics seem positively associated with crime solving, some crimes remain unsolved due to 'incompetence, misfeasance, nonfeasance resource problems, or simple bad luck' of police. In the study by James and Beauregard (2018), 62 cases of SHs were analyzed to understand factors influencing the duration of crime solving. Researchers tested modus operandi characteristics and criminal investigation factors in these cases. Their findings showed that criminal investigation factors are mainly responsible for the duration of the investigation. Specifically, the body discovery by civilians and a body in advanced decomposition increase the length of the investigation, as well as decisions by the investigators that could be considered contentious (i.e., failure to conduct a neighborhood canvas; failure to use a DNA sample from the crime scene; failure to follow up on a missing person's report concerning the victim; failure to investigate the location at which the victim disappeared; failure of investigative teams to share information; drawing conclusions concerning the nature of the victim's death that do not reflect the facts; conviction of an innocent third part) (James and Beauregard 2018).

9 Conclusion

This chapter discussed the various strategies used by sex offenders to avoid police detection. Largely influenced by the rational choice approach, we have presented research findings suggesting an important heterogeneity in the use of FAS by sex offenders. This chapter also introduced findings related to collection and analysis of forensic evidence during the criminal investigation process. These findings stemming from the research on FAS and crime solving techniques over the last 30 years may allow for an improved understanding of sex offenders' behavior, as well as police practices to improve crime solving. On the one hand, the use of FAS by sex offenders appears to have a limited impact on avoiding police detection, while some strategies can be counterproductive and increase the risk of arrest. On the other hand, research findings highlight problems in the collection and treatment of different evidence by law enforcement during the investigative process. Based on these findings, police could adapt investigative strategies by focusing specifically on techniques used by sex offenders that have a sizable and reliable impact on the crime being solved. It appears that the process to detect all FAS used by sex offenders and collect forensic evidence is not an efficient method to solve sex crimes. The attrition process between evidence collection and analysis appears as a waste of time and resources for law enforcement. Forensic evidences are not a guarantee of success for criminal investigations, and must be considered as an additional tool to serve investigators in certain situations, rather than the ultimate tool that will allow to solve all situations. Investigators should develop their skills in alternative ways to collect fundamental information like victims' testimonies using the most appropriate techniques, such as the Cognitive Interview. They should also adapt human resources depending on the complexity of the crime characteristics in crime solving. Future studies should continue to analyze the use of FAS by sex offenders to understand the criminal process and the complexity surrounding rational choice approach in a criminal context. More specifically, research should examine the learning process to avoid police detection of sex offenders from the contact with the criminal justice system. Research should also specifically analyze which evidence has an impact on crime solving for different types of sex crimes (i.e., sexual abuse, rapes, SH) involving different types of victims (i.e., children, teenagers, adult, elderly).

Finally, research should also try to link the modus operandi and FAS in order to identify if offenders using sophisticated modus operandi to commit a rape are also using sophisticated strategies to avoid police detection.

References

Balemba, S., Beauregard, E., & Martineau, M. M. (2014). Getting away with murder: A thematic approach to solved sexual homicides using crime scene factors. *Police Practice and Research: An International Journal, 15*(3), 221–233.

Beauregard, E. (2005). *Processus de prédation des agresseurs sexuels sériels: Une approche du choix rationnel* (PhD), Université de Montréal, Montréal.

Beauregard, E., & Bouchard, M. (2010). Cleaning up your act: Forensic awareness as a detection avoidance strategy. *Journal of Criminal Justice, 38*(6), 1160–1166.

Beauregard, E., & Field, J. (2008). Body disposal patterns of sexual murderers: Implications for offender profiling. *Journal of Police and Criminal Psychology, 23*(2), 81.

Beauregard, E., & Leclerc, B. (2007). An application of the rational choice approach to the offending process of sex offenders: A closer look at the decision-making. *Sexual Abuse, 19*, 115–133.

Beauregard, E., & Martineau, M. M. (2013). A descriptive study of sexual homicide in Canada: Implications for police investigation. *International Journal of Offender Therapy and Comparative Criminology, 57*(12), 1454–1476.

Beauregard, E., & Martineau, M. M. (2014). No body, no crime? The role of forensic awareness in avoiding police detection in cases of sexual homicide. *Journal of Criminal Justice, 42*(2), 213–220.

Beauregard, E., & Martineau, M. M. (2016). Does the organized sexual murderer better delay and avoid detection? *Journal of Interpersonal Violence, 31*(1), 4–25.

Beauregard, E., & Martineau, M. M. (2017). Can sexual murderers avoid police detection? In E. Beauregard & M. M. Martineau (Eds.), *The sexual murderer: Offender behaviour and implications for practice* (Vol. 173–206). New York: Routledge.

Beauregard, E., & Martineau, M. M. (2018). Action-reaction? Forensic evidence analysis and forensic awareness in sexual homicide. In M. DeLisi & M. G. Vaughn (Eds.), *The Routledge international handbook of biosocial criminology* (pp. 605–622). New York: Routledge.

Beauregard, E., Proulx, J., Rossmo, K., Leclerc, B., & Allaire, J. F. (2007a). Script analysis of the hunting process of serial sex offenders. *Criminal Justice and Behavior, 34*(8), 1069–1084. https://doi.org/10.1177/0093854807300851.

Beauregard, E., Rossmo, K., & Proulx, J. (2007b). A descriptive model of the hunting process of serial sex offender: A rational choice perspective. *Journal of Family Violence, 22*, 444–463.

Beauregard, E., Rebocho, M. F., & Rossmo, K. (2010). Target selection patterns in rape. *Journal of Investigative Psychology and Offender Profiling, 7*(2), 137–152.

Beaver, K. M. (2010). The promises and pitfalls of forensic evidence in unsolved crimes. *Criminology & Public Policy, 9*(2), 405–410.

Bouchard, M., & Nguyen, H. (2010). Is it who you know, or how many that counts? Criminal networks and cost avoidance in a sample of young offenders. *Justice Quarterly, 27*(1), 130–158.

Brecklin, L. R., & Ullman, S. E. (2010). The role of victim and offender substance use in sexual assault outcomes. *Journal of Interpersonal Violence, 25*(8), 1503–1522.

Brown, K. M., & Keppel, R. D. (2012). Child abduction murder: The impact of forensic evidence on solvability. *Journal of Forensic Sciences, 57*(2), 353–363.

Chopin, J. (2017). *La gestion des liens entre les crimes sexuels de prédations: Repenser ViCLAS sous la perspective du paradigme situationnel* [Identifying links among predatory sexual crimes: Rethinking ViCLAS under the situational paradigm perspective] (Doctoral thesis), Université de Lausanne, Lausanne.

Chopin, J., & Beauregard, E. (2018). The sexual murderer is a distinct type of offender. *International Journal of Offender Therapy and Comparative Criminology*. https://doi.org/10.1177/0306624X18817445.

Chopin, J., & Beauregard, E. (2019a). Elderly sexual abuse: An examination of the criminal event. *Sexual Abuse: A Journal of Research and Treatement*. https://doi.org/10.1177/1079063219843899.

Chopin, J., & Beauregard, E. (2019b). Sexual Homicide in France and Canada: An International Comparison. *Journal of interpersonal violence*. https://doi.org/10.1177/0886260519875547.

Chopin, J., & Beauregard, E. (2019c). Sexual homicide of children: A new classification. *International Journal of Offender Therapy and Comparative Criminology*. https://doi.org/10.1177/0306624X19834419.

Chopin, J., & Caneppele, S. (2019a). Geocoding child sexual abuse: An explorative analysis on journey to crime and to victimization from French police data. *Child Abuse & Neglect, 91*, 116–130. https://doi.org/10.1016/j.chiabu.2019.03.001.

Chopin, J., & Caneppele, S. (2019b). The mobility crime triangle for sexual offenders and the role of individual and environmental factors. *Sexual Abuse: A Journal of Research and Treatement, 31*(7), 812–836. https://doi.org/10.1177/1079063218784558.

Chopin, J., Beauregard, E., Bitzer, S., & Reale, K. (2019). Rapists' behaviors to avoid police detection. *Journal of Criminal Justice*. https://doi.org/10.1016/j.jcrimjus.2019.04.001.

Ciavaldini, A. (1999). *Psychopathologie des agresseurs sexuels*. Paris: Masson.

Cole, S. A. (2010). Forensic identification evidence: Utility without infallibility. *Criminology & Public Policy, 9*(2), 375–379.

Cole, S. A., & Dioso-Villa, R. (2006). CSI and its effects: Media, juries, and the burden of proof. *New England Law Review, 41*, 435.

Cornish, D. B. (1994). The procedural analysis of offending and its relevance for situational prevention. In R. V. Clarke (Ed.), *Crime prevention studies*. New York: Transaction Press.

Cornish, D. B., & Clarke, R. V. (1986). Introduction. In D. B. Cornish & R. V. Clarke (Eds.), *The reasoning criminal: Rational choice perspectives on offending*. New York: Springer.

Cornish, D. B., & Clarke, R. V. (1987). Understanding crime displacement: An application of rational choice theory. *Criminology, 25*(4), 933–948.

Cromwell, P. F., & Olson, J. N. (2004). *Breaking and entering: Burglars on burglary*. Belmont: Wadsworth.

Davies, A. (1992). Rapists' behaviour: A three aspect model as a basis for analysis and the identification of serial crime. *Forensic Science International, 55*(2), 173–194.

Davies, A., & Dale, A. (1995). *Locating the stranger rapist*. London: Home Office Police Department.

Davies, A., Wittebrood, K., & Jackson, J. L. (1997). Predicting the criminal antecedents of a stranger rapist from his offence behaviour. *Science & justice: journal of the Forensic Science Society, 37*(3), 161–170.

Du Mont, J., & Parnis, D. (2000). Sexual assault and legal resolution: Querying the medical collection of forensic evidence. *Medecine & Law, 19*, 779–792.

Fisher, B. A. J. (2003). *Techniques of crime scene investigation*. Boca Raton: CRC press.

Hazelwood, R. R., & Burgess, A. W. (2001). The behavioral oriented interview of rape victims: The key to profiling. In R. R. Hazelwood & A. W. Burgess (Eds.), *Practical aspects of rape investigation: A multidisciplinary approach* (pp. 115–131). Boca Raton: CRC Press.

James, J., & Beauregard, E. (2018). Murderer vs investigator: Factors influencing the resolution of sexual homicide cases. *Police Practice and Research*, 1–15. https://doi.org/10.1080/15614263.2018.1526683.

James, S. H., & Nordby, J. J. (2005). *Forensic science: An introduction to scientific and investigative techniques* (2nd ed.). Boca Raton: CRC Press.

Johnson, D., Peterson, J., Sommers, I., & Baskin, D. (2012). Use of forensic science in investigating crimes of sexual violence: Contrasting its theoretical potential with empirical realities. *Violence Against Women, 18*(2), 193–222.

Locard, E. (1920). *L'enquête criminelle et les méthodes scientifiques [Criminal investigation and scientific methods]*. Paris: Flammarion.

Lussier, P., & Mathesius, J. (2012). Criminal achievement, criminal career initiation, and detection avoidance: The onset of successful sex offending. *Journal of Crime and Justice, 35*(3), 376–394.

Lussier, P., Bouchard, M., & Beauregard, E. (2011). Patterns of criminal achievement in sexual offending: Unravelling the "successful" sex offender. *Journal of Criminal Justice, 39*(5), 433–444.

McGregor, M. J., Du Mont, J., & Myhr, T. L. (2002). Sexual assault forensic medical examination: Is evidence related to successful prosecution? *Annals of Emergency Medicine, 39*(6), 639–647.

Morselli, C., & Royer, M.-N. (2008). Criminal mobility and criminal achievement. *Journal of Research in Crime and Delinquency, 45*(1), 4–21.

Morselli, C., & Tremblay, P. (2004). Criminal achievement, offender networks and the benefits of low self-control. *Criminology, 42*(3), 773–804.

Morselli, C., Tremblay, P., & McCarthy, B. (2006). Mentors and criminal achievement. *Criminology, 44*(1), 17–43.

Murray, J. M. (2007). *Situationnal factors in rape* (PhD), Griffith University, Griffith.

Nguyen, H., & Bouchard, M. (2013). Need, connections, or competence? Criminal achievement among adolescent offenders. *Justice Quarterly, 30*(1), 44–83.

Park, J., Schlesinger, L. B., Pinizzotto, A. J., & Davis, E. F. (2008). Serial and single-victim rapists: Differences in crime-scene violence, interpersonal involvement, and criminal sophistication. *Behavioral Sciences & the Law, 26*(2), 227–237.

Peterson, J. L., Mihajlovic, S., & Gilliland, M. (1984). *Forensic evidence and the police*. Washington, DC: National Institute of Justice.

Polaschek, D. L., & Ward, T. (2002). The implicit theories of potential rapists: What our questionnaires tell us. *Aggression and Violent Behavior, 7*(4), 385–406.

Reale, K., Beauregard, E., & Martineau, M. M. (2017). Is investigative awareness a distinctive feature of sexual sadism? *Journal of Interpersonal Violence*. https://doi.org/10.1177/0886260517698824.

Rossmo, K. (2009). *Criminal investigative failures*. Boca Raton: CRC Press.

Strom, K. J., & Hickman, M. J. (2010). Unanalyzed evidence in law-enforcement agencies: A national examination of forensic processing in police departments. *Criminology & Public Policy, 9*(2), 381–404.

Testa, M., & Livingston, J. A. (2000). Alcohol and sexual aggression: Reciprocal relationships over time in a sample of high-risk women. *Journal of Interpersonal Violence, 15*(4), 413–427.

Tremblay, P., & Morselli, C. (2000). Patterns in criminal achievement: Wilson and Abrahamse revisited. *Criminology, 38*(2), 633–657.

Ward, T. (2000). Sexual offenders' cognitive distortions as implicit theories. *Aggression and Violent Behavior, 5*(5), 491–507.

Zawacki, T., Abbey, A., Buck, P., McAuslan, P., & Clinton, A. (2003). Perpetrators of alcohol-involved sexual assaults: How do they differ from other sexual assault perpetrators and nonperpetrators. *Agressive Behavior, 29*, 366–380.

Tracking Terrorism: The Role of Technology in Risk Assessment and Monitoring of Terrorist Offenders

Neil Shortland and James J. F. Forest

Abstract Intelligence is a critical aspect of countering violent extremism, and technology is increasingly used to help expand the amount of information collected about potential terrorist threats. However, it is clear that an equally important (if not more important) part of these efforts is the ability to separate, within the mass of individuals (or data) identified, those individuals who will conduct acts of terrorism (the minority) from those who will not (the majority). Thus, a core challenge is how information is handled, interpreted, analyzed, and used to inform decisions. In this chapter we discuss the dynamic interplay between the collection of intelligence (including technology-assisted surveillance) and decision-making. Specifically, we focus on how issues in our understanding of "the terrorist" interact with and impact the use of technology within counterterrorism. In addition to this, and in support of future research in this area, we highlight some innovative areas for growth and new avenues to facilitate the integration of technology within counterterrorism.

Keywords Terrorism · Risk assessment · Intelligence · Counter violent extremism

1 Introduction

For several decades, an increasingly broad range of technology has been used in efforts to identify, track and apprehend individuals engaged in terrorist activity. By various assessment measures, this has contributed to an increased effectiveness in counterterrorism efforts worldwide. And yet, the statement "He [the perpetrator of the attack] was known to authorities" is often heard in the aftermath of incidents of extremist violence. A cursory look at recent cases supports this assessment. Michael Adebowale and Michael Adebolajo were known to intelligence agencies prior to murdering British fusilier Lee Rigby (Intelligence and Security Committee of

N. Shortland · J. J. F. Forest (✉)
University of Massachusetts Lowell, Lowell, MA, USA
e-mail: Neil_Shortland@uml.edu; james_forest@uml.edu

B. Fox et al. (eds.), *Science Informed Policing*, Advanced Sciences and Technologies for Security Applications,
https://doi.org/10.1007/978-3-030-41287-6_4

Parliament 2014). Nidal Malik Hassan's emails had been analyzed by the FBI prior to his attack at Ft. Hood, Texas on February 8, 2009 (Webster Commission 2012). Man Haron Monis was assessed as a risk the day before he held siege to a Lindt coffee store in Sydney, Australia on December 15–16, 2014 (Joint Commonwealth NSW Review 2015). Tamerlan Tsarnaev (the architect of the April 15, 2013 Boston Marathon attack) was known to authorities, as were the perpetrators of terrorist attacks in Paris (November 13, 2015), London (July 7, 2005), Madrid (March 11, 2004), New York and Washington, DC (September 11, 2001), and many others (Gallagher 2015). Thus, despite improved data collection methods, improved technologies, decades of research and constant innovations around the science of counterterrorism, there remains much more work to be done to improve the effectiveness of counterterrorism—and specifically developing technologies that offer promising new avenues to pursue. In this chapter we will outline both the use (and future) of technology in counterterrorism, as well as the practical "ground truths" of what countering terrorism requires. As we shall outline, while there is a lot that we can ask of technology, and a mass of services it can provide, there are some factors within the phenomenon of "terrorism" (low-base rates, low signals, high signal-to-noise ratios, added issues of false flags due to the nature of Internet consumption) that continue to pose problems for any technological solution.

Before proceeding, however, we must first establish what we mean by the terms terrorism and counterterrorism, as neither has a universally accepted definition. First, terrorism is generally viewed as a type of deviant behavior that intersects criminality, political goals and communication. Terrorists consciously choose to kill, maim and destroy, and also routinely engage in money laundering, theft, fraud, extortion, hijacking, smuggling (including drugs, weapons, and humans), kidnapping, bank robbery and many other kinds of criminal activity. The motivations behind such activity are largely described by a type of ideology, an articulation of sociopolitical grievances and goals. The Department of Justice defines terrorism as "the unlawful use of force or violence against persons or property to intimidate or coerce a government, the civilian population, or any segment thereof, in furtherance of political or social objectives" (Bjelopera 2017). Further, there are many different ideological categories that can describe the sociopolitical motivations behind terrorist activity, including left-wing, ethno-nationalist, religious, right-wing, environmental extremist, anti-abortion, etc. And unlike many other violent criminals (who tend to avoid the media), terrorists often claim responsibility for attacks and try to justify their violence through their ideology, an articulation of their grievances and goals. This communicative aspect is highlighted by Bruce Hoffman's (2006, p. 40–41) definition of terrorism as "the deliberate creation and exploitation of fear through violence or the threat of violence in the pursuit of political change … [it is] designed to have far-reaching psychological effects beyond the immediate victim(s) or object of the terrorist attack … [and] to create power where there is none or to consolidate power where there is very little."

The term counterterrorism is generally used to describe the various means by which governments and communities seek to turn people away from supporting or engaging in terrorist activity of any kind. Most countries employ a range of legal instruments toward this goal, conveying a message to the populace that those individuals who engage in terrorism will be punished to the full extent of the law. Throughout the twentieth century, counterterrorism efforts in the U.S. were mainly the responsibility of law enforcement, because terrorist acts (of both foreign and domestic origin) against American targets were viewed as crimes to be investigated and prosecuted under U.S. law. Law enforcement agencies—and in particular, the Federal Bureau of Investigation (FBI)—mounted aggressive investigations of suspected terrorists, expanded the use of informants, and built a stronger intelligence base (Rosenau 2014, p. iii). During the 1970s and 1980s, a series of presidential determinations on counterterrorism gave the U.S. State Department the lead role overseas, while the FBI would focus its efforts domestically, and the Federal Aviation Administration would be in charge when responding to airplane hijackings (Kraft and Marks 2012, p. 328).

These responsibilities remained largely unchanged for the next four decades, until the terrorist attacks of September 11, 2001. Since then, according to their own analysis, the FBI has: (1) shifted its counterterrorism culture and organization from reactive to proactive and "threat-based;" (2) developed a nationally-driven, fully integrated Intelligence and Investigative Program; (3) improved information sharing with other federal agencies, state and local governments, and international counterterrorism partners; (4) enhanced operational capabilities within FBI Headquarters and the field; and (5) evaluated lessons learned to better equip the nation in preventing terrorism (FBI 2004). A report by the International Association of Chiefs of Police describes how, after the attacks of 9/11, law enforcement agencies in the U.S. were "asked to confront a new threat to their communities, perpetrated by individuals and organizations that had vastly different motivations and means of attack from traditional criminals. Accepting this challenge required law enforcement agencies to reassess their operations and reevaluate their priorities" (International Association of Chiefs of Police 2005). However, according to Chief Deputy Jose Docobo of the Hillsborough County Sheriff's Office in Tampa Florida, it has become clear that "many of the objectives of terrorism prevention parallel current law enforcement policies with respect to local crime issues" (Docobo 2006).

In his analysis of the intersection between terrorism, homeland security and community policing initiatives, Docobo observes that: "the investigative approach to a terrorist event is similar to that of a traditional crime incident. Creating external partnerships, citizen involvement, problem solving, and transforming the organization to take on a new mission are all key elements of community policing which must also be part of a comprehensive homeland policing strategy. Community policing, which builds trust between the community and law enforcement and gives officers knowledge of the community and resident activity, is essential to providing actionable intelligence relating to potential terrorist actions" (Ibid). The acquisition and analysis of this intelligence, as we will discuss later in this chapter, forms the cornerstone of any successful counterterrorism effort.

However, there are also several other dimensions of counterterrorism beyond legal/law enforcement and intelligence that must be recognized. For example, through the Financial Crimes Enforcement Network, banks and other institutions can monitor suspicious transactions and freeze the assets of suspected terrorist groups or individuals. In cases of terrorism originating from foreign countries, counterterrorism may involve diplomats and the military.[1] A nation can use economic policies as a form of leverage to coerce other nations into curbing state sponsorship of terrorist groups. Counterterrorism may also employ public diplomacy and psychological operations. The integration of all these and other dimensions is critical to the success of any counterterrorism effort (Forest 2015). Further, this complex terrain of efforts requires the involvement of many government agencies and personnel: border patrols, airport security, maritime port cargo screening machines, immigration services, law enforcement and corrections officers, finance and treasury departments, military and intelligence services, and even public health agencies participate in some aspect of counterterrorism.

In sum, counterterrorism is, as Alex Schmid (2013, p. 620) explains, "a proactive effort to prevent, deter, and combat politically motivated violence directed at civilian and non-combatant targets, by the use of a broad spectrum of response measures." Because terrorism is a form of violent human behavior, the success or failure of counterterrorism efforts can largely depend on the quality of intelligence about an individual's intentions and capabilities. Questions driving the collection and analysis of intelligence include "Who is a terrorist? Who will carry out the next terrorist attack, and where are they now? What are they planning to do? When, where and why?" Authorities have increasingly turned to various kinds of technology to address these questions, as we'll describe in this chapter. Much of this technology has been employed with a "more is better" mentality, where massive amounts of data are collected from an increasingly broad spectrum of sources. As a result, a lack of information is generally not the primary challenge—rather, it is how that information is processed, analyzed, and converted into operationally useful intelligence.

What is needed now is the ability to identify individuals who intend to conduct acts of terrorism (a very small percentage of a population) among those who do not (the overwhelming majority). To this end, simulations, massive databases and advanced algorithms are being used by law enforcement and other government agencies to make decisions and improve the effectiveness of counterterrorism efforts. After reviewing several examples of how technology assists counterterrorism efforts in general, this chapter will focus on the need to effectively prioritize and analyze the intelligence collected in ways that reduce the "noise to signal" ratio. Then in the latter part of the chapter, we will explore some new and emerging technologies—like automatic text detection and machine learning—that offer promising potential for improving counterterrorism. And finally, the discussion will conclude with some thoughts on implications for law enforcement, federal research and development, and the role of the private sector in creating these new technologies.

[1] In the U.S., the role of the military in counterterrorism is limited to overseas activities, but this is not the case in many other countries

2 Technologies and Counterterrorism

Each year the U.S. federal budget supports hundreds of technology research and development projects that can yield benefits for counterterrorism. Some projects focus on identifying explosive materials, or providing explosive ordnance disposal technicians (bomb squads) with new tools to safely disarm or destroy those weapons. For example, a van was developed with special equipment that allows the driver to surreptitiously drive past other vehicles or cargo containers and scan for explosives and other contraband. Other projects address the threat of terrorists acquiring and using chemical, biological, radiological, or nuclear materials. Examples include sensors in public places that can detect harmful chemicals or airborne pathogens, or a lightweight mask that will provide up to 20 minutes of protection against biological and chemical agents and can be carried in a briefcase or large coat pocket (Kraft 2015). The Department of Agriculture uses a range of technologies to mitigate the chances that terrorists could successfully attack the food supply, perhaps through poisoning or contaminating food before it reaches supermarket shelves.

New technologies are also being developed to improve airport security checkpoint and baggage screening, while improved communication systems help ensure law enforcement, medical, fire and rescue personnel can coordinate with each other (one of several failings identified in the *9/11 Commission Report*). And while physical security and personal protection are prominent goals of these technologies, the Department of Justice has become a global leader in the field of forensic science research and technology development. In particular, the FBI's Counterterrorism and Forensic Science Research Unit (CFSRU) has helped improve the investigative capabilities of numerous federal, state and local law enforcement agencies.

In short, technologies are used for many different aspects of counterterrorism. But perhaps the most important from the standpoint of preventing terrorist attacks involves the collection and analysis of information that can be used to identify and thwart individuals with the intentions and capabilities for such an attack. In fact, given the primacy of intelligence in the overall mission of counterterrorism, it makes sense that so much technology is used for these purposes.

Surveillance of suspects is perhaps the most well-known application of technology in counterterrorism. Audio recording (eavesdropping, wiretapping, etc.) has been a part of law enforcement investigations since the invention of the telephone. Today, systems are capable of listening to live conversations and automatically triggering a recording when a particular word or phrase is spoken. And while human translators are still highly valued, computerized language translators and transcribers have been developed with increasingly sophisticated capabilities. These surveillance efforts have not been without some controversy. For example, in December, 2005 the *New York Times* published an article exposing the fact that the Bush administration had "secretly authorized the National Security Agency to eavesdrop on Americans and others inside the United States to search for evidence of terrorist activity without the court-approved warrants ordinarily required for domestic spying … Under a presidential order signed in 2002, the intelligence agency has monitored the international telephone calls and international e-mail messages of hundreds, perhaps thousands, of people inside the United States without warrants over the past three years in an effort to track possible 'dirty numbers' linked to Al Qaeda" (Risen and Lichtblau 2005), although the agency still seeks warrants to monitor entirely domestic communications.

Meanwhile, video surveillance has seen many advances as well, from new closed-circuit television systems and night vision cameras to satellites that can read a license plate from space. Unmanned aerial vehicles, originally developed for the military, have been increasingly used by U.S. border security and by law enforcement for kidnapping investigations, search and rescue operations, drug interdictions and fugitive investigations (Department of Justice n.d.; Friedersdorf 2016). Notably, the Department of Justice has issued guidance for the legal use of these drones, noting that they "must be operated consistent with the U.S. Constitution… and may only be used in connection with properly authorized investigations and activities" (Ibid).

Signals intelligence is another key area of intelligence collection. Law enforcement can partner with cellular phone service providers to track users when their phone identifies its location to nearby cell towers. According to one analysis, the precision of cell site location information (CSLI) can be made to calculate a few feet from where the person is standing (Bloom and Clark 2016). Miniaturized transponders attached to a suspect's vehicle, like what we use to pay tolls on the highway, can also reveal their location and travel direction to authorities seeking his or her whereabouts. Credit card transactions can be obtained and studied to detect any suspicious purchases or travel arrangements. A terror suspect's digital activities can be closely monitored as well. E-mail and other electronic communications can be easily intercepted and archived, web browsing history can be searched, and key logging tools can record what is typed on a keyboard. Social media posts, likes, shares and other indications of personal beliefs can be collected and analyzed.

In truth, there is no shortage of information that can be gathered. Thus, the drive to "get more information" combined with the increasing ease by which huge amounts of information can be obtained led to the creation of several ambitious government efforts over the past few decades. For example, the Section 702 Surveillance Program was originally approved by Congress in 2008 to increase the government's ability to track and thwart foreign terrorists in the wake of the 9/11 attacks, and according to leaders of the U.S. intelligence community it has helped thwart several terrorist plots against the U.S., including a planned attack on the New York City subway system (Kelly 2018). With alleged code names like Echelon, Carnivore, PRISM, XKeyscore, Tempora, Muscular, Marina, Mainway, Pinwale, Project6, Storeroom, Lustre, NarusInsight and others, these surveillance programs often tried to simply capture as much information as possible (from multiple sources) and then sift through it looking for keywords, phrases, patterns or anything that could be deemed "suspicious" (BBC News 2001).

But like the audio surveillance controversy mentioned earlier, massive database collection projects by the government have also been the source of heated public debate about whether Americans are trading security for reductions in privacy and civil rights. For example, in June 2013, Edward Snowden—a contractor for the National Security Agency—stole a trove of classified documents and gave them to reporters. Among these documents were several that revealed previously secret court orders requiring telecommunication firms to provide a range of records and other information to authorities. This touched off a media firestorm, the legacy of which still invokes fear among many Americans about invasive government agents peering over their shoulder (Matney 2015). However, as many practitioners and researchers have acknowledged, having access to information about potential threats to homeland security is not really the core challenge. Rather, the much more difficult task is sifting through tons of information in a way that accurately identifies a person (or persons) about whom authorities should be concerned.

Let us consider a real world example: when Michael Adebolajo communicates with a known member of al-Qaeda in the Arabian Peninsula (AQAP) and says he wishes to become a "martyr", these comments could be viewed by intelligence analysts as indicative of a potentially threatening individual, or they could be viewed as "a fairly standard example of [online] rhetoric" (Intelligence and Security Committee Report 2014, p.120). Here then, there are many people who would (likely) be flagged as "potential threats" based on their use of online rhetoric that emphasizes the need for Jihadist violence. But the issue is that (given the success of technology-based identification solutions) the number of people identified would still exceed the number of resources available to properly mange (i.e., observe and assess) this group. This means that a secondary prioritization is required *after* the identification of the individuals. It is because of issues in this stage that we often hear "the offender was known to authorities".

3 The Internet and Counter Terrorism

At the intersection of modern technology and the threat of terrorism, we find the Internet has become a significant source of valuable information. In a sense, the Internet provides a great opportunity for the use of technology in counterterrorism efforts, but at the same time the Internet also creates a psychological environment that complicates the process of counterterrorism. Scholars and practitioners have increasingly focused on the role of the Internet in terrorism (and counterterrorism) (e.g., Khader et al. 2016), but one area that has been frequently overlooked is how the Internet has psychological effects on an individual's behavior. This psychological dimension has direct implications for how information collected via the Internet should be used for counterterrorism.

It is widely known that with most forms of behavior, the environment influences how we act. In general, we know that heat affects aggression, as does hunger. So too does the Internet. Researchers have long noted that, when online, individuals act in a way that is (to varying degrees) not reflective of the types of behavior they would engage in offline. Individuals "loosen up, feel less restrained, and express themselves more openly" (Suler 2004, p. 321). This phenomenon is referred to as the "online disinhibition effect", and it can create both positive behaviors (e.g., such as disclosing sensitive personal information), and negative behaviors (darker behaviors such as expressing racism and threats against others, which can be described as "toxic" disinhibition; see Postmes et al. 2002).

When an individual is on the Internet, several factors contribute to this sense of lowered inhibitions (see Suler 2004). One of the most central factors is anonymity, in that when we are online there is a sense (perceived at least) that our identity is hidden. While anonymity has clear counter-security benefits (see Weimann 2004, 2006) it has profound psychological effects on how we behave (Zimbardo 1969). Anonymity reduces self-awareness and "accountability cues," in turn lowering the degree to which we abide by our own internal standards (Prentice-Dunn and Rogers 1989). At the same time, the Internet removes real-world social restraints. It allows people to pretend to be someone they are not. This effect is known as the "equalization hypothesis" (see Dubrovsky et al. 1991). As Gelder (2006, p. 37) argues, this means that the "vulnerable border between fantasy and reality" becomes even more vulnerable. As Suler (2004, p. 325) observes:

> The disinhibition effect can then be understood as the person shifting, while online, to an intrapsychic constellation that may be, in varying degrees, dissociated from the in-person constellation, with inhibiting guilt, anxiety, and related affects as features of the in-person self but not as part of that online self. This constellations model—which is consistent with current clinical theories regarding dissociation and information processing—helps ex- plain the disinhibition effect as well as other online phenomena, like identity experimentation, role playing, multitasking, and other more subtle shifts in personality expression as someone moves from one online environment to another. In fact, a single disinhibited 'online self' probably does not exist at all, but rather a collection of slightly different constellations of affect, memory, and thought that surface in and interact with different types of online environments.

This research on the psychological dimensions of online behavior has direct implications for using technology to assist in counterterrorism in that people are likely to behave in ways, or express intentions to behave, that are not directly correlated with their actual intentions. It is for this reason that expressing the desire to become a martyr is "a fairly standard example of [online] rhetoric" (Intelligence and Security Committee Report 2014, p.120).

Related to this issue then is the sheer volume of extremist material on the Internet; and associated with this, the sheer number of people engaging with such content. Brachman (2009) notes that websites, blogs and message boards advocating violent extremism grew exponentially in the aftermath of 9/11. Furthermore, such material garners a significant amount of "traffic." The beheading of Nick Berg (conducted, filmed and posted online by Islamic extremists) was downloaded over 15 million times, and sites that hosted the video received over 60,000 hits per hour (Talbot 2005, p. 2). Berger and Morgan (2015) reported that between September and December 2014, there were 46,000 Twitter accounts linked to ISIS supporters (a conservative estimate, with an upper limit of 70,000; p. 2). As we will highlight below, engagement in terrorism is viewed as a low-base rate activity. This means that the proliferation of extremist related content online, and proliferation of extremist "intent" is merely increasing the amount of "noise" in the system, compounding the difficulty of pinpointing the signals we need to see in order to identify individuals that require greater attention and preventive measures.

One final problem worth mentioning is the difference between the kinds of information available and (crucially) information that is not available online. This is especially true when individuals participate in closed networks, as well as on platforms owned by third-parties (such as Facebook). This point is well demonstrated in the risk assessment process for Michael Adebowale and Michael Adebolajo. Michael Adebowale and Michael Adebolajo were convicted in December 2013 for the murder of fusilier Lee Rigby, but prior to this both offenders were known to the domestic security services (Intelligence and Security Committee Report 2014). In the ensuing review of the incident, the following information was identified:

> …a substantial online exchange between Adebowale and FOXTROT in December 2012, in which Adebowale expressed his desire to murder a soldier – in the most graphic and emotive manner – because of UK military action in Iraq and Afghanistan. Adebowale had not, at that point (five months before the attack), developed a definite plan as to how he might carry out such an attack. FOXTROT encouraged him and suggested several potential attack methodologies, ranging from a martyrdom operation to use of a knife. Adebowale believed that security arrangements that guarded soldiers' places of work might make it difficult to carry out an attack, and that alternative, less secure locations should be considered

It was stated in the review that had this exchange been seen, "Adebowale would have become a top priority… [and] there is a significant possibility that MI5 would then have been able to prevent the attack." However, "there may have been only a very slim chance that MI5 would have had sight of the FOXTROT exchange" (Intelligence and Security Committee Report 2014, p. 7).

4 Threat Management vs. Threat Identification

The good news is that a broad range of technologies can be used to gather a significant amount of information about many potential suspects. Yet at the same time, the bad news is that the amount of information is so significant, it risks overwhelming the analytical capabilities of those doing the collecting. In May 2018, the National Security Agency reported that it had collected more than 534 million records of phone calls and text messages from American telecommunications providers like AT&T and Verizon during the previous year—more than three times what it collected in 2016 (Office of the Director of National Intelligence 2018). Thus, the problem for law enforcement investigations into potential terrorist activity becomes one of threat information management and not threat identification.

When thinking about the role of technology in counterterrorism we have to think about moving beyond identification (i.e., identifying who may potentially pose a threat) and towards efforts to assist prioritization (i.e., identifying who, among many, potentially poses *the most severe* threat). While identification is important, prioritization will allow enforcement agencies to use their finite resources in a way that best protects the public.

Now, the issue of threat prioritization (either with or without technology) is not new, and in fact some of the earliest efforts in the study of the psychology of "the terrorist" attempted to identify factors that would allow us to identify what made a terrorist unique. In order for technology to assist in counterterrorism we have to be able to process the information in the correct way; i.e., we need to know what bits of information are most important (in this context, most *indicative* of what a person will or will not do next). This led to the (unsuccessful) quest to identify the "profile" of "the terrorist" since the early 1980s (see Horgan 2013). Early attempts to risk assess violent extremists rested on the assumption that they were "special" or different from members of the general population (i.e., those who would not engage in extremist violence), and that single factors could differentiate "risky" individuals. Some researchers sought to identify psychological "markers"—these included specific manifestations of a mental illness like narcissism or psychopathy or suicidal tendencies (e.g., Lankford 2013)—while others examined demographic differences (Sageman 2004). Overall, research in this field has generally concluded that people who become involved in terrorism are demographically unremarkable (see Bakker 2006; Sageman 2004; Simcox and Dyer 2013) and there were no specific mental illnesses that defined them (though for updated work on this see Corner and Gill 2015). In fact, most researchers have instead begun to focus on differences that exist between different types of actor (or different ideologies of actor) as a more fruitful area of study, compared to looking for differences between "terrorists" and "non-terrorists" (Gill 2013; Gruenewald et al. 2013a, b; Horgan et al. 2016).

An alternate area of study was to identify (or apply) risk assessment tools that brought together several different factors that would, in combination, be indicative of extremist intent (and thus a useful and effective way of prioritizing threats). There have been several efforts in this domain to identify what factors are the most important in predicting the likelihood that an individual will engage in an act of terrorism. Kebbell and Porter (2011) conducted one of the earliest analyses in this area and identified a series of psychological factors that would be associated with those who were most likely to engage in violent extremism. They listed these factors in accordance with four over-arching sub-categories: "standard risk factors" (high frequency, with low predictive utility, such as a super-ordinate non-Western identity or a negative view of Western foreign policy); "moderate risk factors" (indicative of higher risk and more supportive of violence, such as religious beliefs that support violence); "higher risk factors" (those associated with enhanced operational capability, such as membership in a radical political group) and "extreme risk factors" (those that show an individual has decided to engage in violent extremism). The Violent Extremist Risk Assessment (VERA) protocol consists of 31 indicator items that cover beliefs, attitudes, intent, history, capability, commitment and motivation (along with some protective items; Pressman and Flockton 2014). VERA-2 is used as part of a multi-modal risk assessment that includes general violence risk and provides "offense-specific supportive information on the risk of future violence of convicted and incarcerated violent extremist offenders" (Pressman and Flockton 2014, p. 138). More recently, the CYBERA tool (see Pressman and Ivan 2016) was developed to be used alongside the VERA-2 to provide "a robust and cyber-focused risk assessment intended to provide early warning indicators of violent extremist action" (p. 391). The CYBERA tool includes a series of indicators focused on online behavior (imagery, semantic content, beliefs and intentions expressed online and network connectivity) as well as a scale of intensity for each indicator (low-moderate-high). Finally, Meloy (2011; Meloy and Gill 2016) has incorporated risk assessment models that focus on both proximal warning behaviors for targeted violence (Meloy 2011) and distal characteristics of the lone terrorist (Meloy and Yakeley 2014) into the Terrorist Risk Assessment Protocol 18 (TRAP-18).[2]

In principal, any of these tools could be used with technology to be incredibly effective technology-based solutions. But the issue is that, as of this writing, there are several reasons hindering the effectiveness of these risk assessment tools. Recently, Gill (2015a) outlined eight main problems with the field. The first issue is the sheer number of "indicators" that could be used to identify "risk". Gill et al. (2014) for example identified over 100 potential factors. The second issue (linked to the first) is base rates—we are currently unaware of the base rates of these indicators in non-violent extremist samples and non-violent non-extremist samples. What this means is that while Meloy and Gill (2016) demonstrate the prevalence of TRAP-18 factors in a sample of lone actors, this may not have any predictive utility if such factors are also equally prevalent in a sample of non-lone actors. In such cases, research is increasingly (and worryingly) sampling on the dependent variable (Gill 2015a, b).

The third and fourth issues include a lack of understanding (and integration) of protective factors and factor weighting (i.e., the relative importance of one factor over another). Fifth, we also have little understanding of behavioral clustering (i.e., the co-occurrence of behaviors), although recent work using smallest-space analyses is attempting to overcome this (see Horgan et al. 2018). The sixth issue is definitional, and the singular attempt to risk assess "the terrorist" which ignores the known role (and ideology) specific nature of offenders (see; Gill et al. 2014; Horgan et al. 2016). The seventh and eighth issues both focus on time—both the occurrence of risk factors within time (i.e., what is the effect of a given factor coming first or last in an individual's sequence) and that the study of low-base rate offenders often ignores temporal aspects that may be unique to their cohort. It is also worth adding that due to a lack of sufficient data and empirical research, we have very little insight into the interaction of risk assessment and information uncertainty.

Similar to Gill (2015a), Sarma (2017) also highlighted several challenges that have previously hindered the field of risk assessment. The first issue is risk (hazard) specificity, defined as the need for specificity in what, specifically, is being risk assessed. Risk models do not perform equally well across individuals who engage

[2]The TRAP-18 Manual and Code Sheets can be found online at: https://www.gifrinc.com/trap-18-manual/

in varied forms of behavior, even when these behaviors are subsets of the same wider crime type (e.g., Fernandes et al. 2010). It has long been lamented that scholars too often view terrorist offenders as "binary" (they are, or they are not; cf. Horgan et al. 2016). The second issue involves identifying both risk and resilience factors. As outlined by Sarma (2017), a "difficulty for those completing risk assessments for terrorism is the lack of authoritative evidence on the risk and resilience factors that influence the individual's future risk propensity" (p. 282). The third challenge is screening out true negatives. As noted earlier, risk assessment tools often demonstrate their prevalence in the sample of interest, without demonstrating the lack-of-relevance of given risk factors for the sample that is not of interest. The final problem highlighted by Sarma (2017) is the oft-cited "low base rate problem" described earlier, which compounds the sensitivity and specificity issues (see also Roberts and Horgan 2008). As Sarma (2017) observes, "If the base rate of terrorism drops to 10% of the population, the overall accuracy of the tool decreases to one in four. Given that the true base rate for 'involvement in terrorism' in the West is negligible, on a purely statistical level, actuarial risk assessment systems cannot work" (p. 283).

What this means then, is that despite technology assisting the collection of large numbers of variables about a large number of potentially risky individuals, and there being a host of potentially important variables being identified, we are currently unable to identify the correct way in which these variables need to be analyzed, or combined, in a way that effectively and reliably allows prioritization judgements to be made. This is not factoring in the issue that risk is also temporal and will, therefore, change as dynamic risk factors (such as stressors) will also change over time.

5 Future Possibilities for Technology and Counter terrorism

When considering the role of technology for counterterrorism it is important to highlight a recent argument within the "field" of "terrorism studies" about the quality and progress of the science. In his review of the field, Sageman (2014) writes:

> Despite over a decade of government funding and thousands of newcomers to the field of [extremist] research, we are no closer to answering the simple question of 'What leads a person to turn to political violence?' The state of stagnation with respect to this issue is partly due to the government strategy of funding research without sharing the necessary primary source information with academia, which has created an unbridgeable gap between academia and the intelligence community. This has led to an explosion of speculations with little empirical grounding in academia, which has the methodological skills but lacks data for a major breakthrough. Most of the advances in the field have come from historical archival research and analysis of a few field interviews. Nor has the intelligence community been able to achieve any breakthrough because of the structure and dynamic of this community and its lack of methodological rigor. (p. 565).

His argument is that the field, which is admittedly fast moving and well-funded given societal interest in acts of terrorism and the devastating cost and consequences of acts of terrorism, has failed to make progress because of poor methods, poor data, and poor conceptualization of the problem. While the article itself drew several responses on the positives and progress within the field (e.g., Schmid 2014; Stern 2014; Taylor 2014), recent analyses do demonstrate a relatively low quality of research in the field. Lum et al. (2006) found that only 3% of the 6041 peer-reviewed articles on terrorism (between 1971 and 2003) were based on empirical data. Silke (2007) observed similar findings in his review of the field. While a recent review of the field (2007–2016) suggests this seems to be improving (Schuurman 2018), it is well known that those who study extremism (especially from a psychological perspective) lack the empirical data required to do so.

As such, while there are many papers (largely theoretical) that posit the complexities of the "radicalization" process and how an individual moves from one end of the "radicalized" spectrum to the other (and hence, somehow, closer to violence), in the words of Sageman (2014), "we still don't know what leads people to turn to political violence" (p. 576). In this sense, as we saw above, the use of technology to assist in supporting counterterrorism decision-making will also find itself hampered by the inability to answer the core question of who turns to extremism (and why).

That said, there may be other, more innovative, ways that technology can assist in our attempts to predict who will (or who will not) engage in acts of violent extremism. Below we present two posts made publicly by individuals who were expressing a grievance on social media.

Post 1:

My people of Facebook!…

I urge everyone on this site to read and listen to what I am about to tell everyone. This farce of a leadership that has been put in place in this country needs to be destroyed! We the people of this United States needs to wake the fuck up to the FACT that GOD just protected ALL of the U.S. people and it is a documented fact that he did so! This leadership of ours was going to implement an EMP over this country in conjunction with the pole reversal that happens every 11 years. This pole reversal has happened every 11 years for millennium!!! it has never failed!

This past September it did! God didn't let the poles switch in September! This is why! Obama and his rogue cabinet ORDERED top officials before September to move 3 nuclear devices to North Carolina to be used for this EMP event AGAINST the AMERICAN people to destroy this country and enslave us. Now, This is the tip of a very touchy subject! How does the American people feel about this little piece of history in the making! All of this that I am telling you is FACT!

Do the research! Its out there! All of you may say that I am off of my rocker and smoking CRACK! I wish the hell I was! This is the truth. Obama and his rogue outfit Ordered this from the Oval Office! He has been put into place here from a long and hard road that the Islamic Fundamentalists have taken! Its here. They have made the American people into laughing stocks throughout this globe of ours!

The American people need to put down their IPADS and wake the fuck up! Look at what is happening here. All of us are in this together! I pray that this doesn't fall on deaf ears! I am taking a big risk putting this truth out for my people to hear. But I love you all! So I am taking that risk! The commanders that were relieved of their duties that foiled this EMP event also put their lives on the line for the American people!

We need to stand up to our Government and take this country back so that our children can have a good FREE life without all of this bullshit that the greed of the world has produced! Its in the hands of the American people now!

Lets see who has the balls to research what I have told you and make the decisions based on these FACTS! By the way, CNN and the media in this country are bought and payed for by the very same people that want to destroy this great land of ours. We are ALL being lied to by this rogue government of ours! We need to stand up against all of this tyranny. GOD BLESS AMERICA!

Post 2:

Hello, my name is [ANONYMIZED] and I support Ed and Elains Brown. Show us a law that requires an American citizen working in the U.S to pay an income tax. Title 26, AKA the internal revenue code, only lists those involved in foreign commerce to pay an income tax, not domestic. That is very important, so where's the law that requires an American citizen working in the U.S to pay an income tax, in fact where's the law that requires an American citizen working in the U.S to file an income tax return? So until I see that law, this is what im gonna do. I have here a standard 1040 U.S individual income tax return (rips). I will not file nor will I pay a single penny in income tax until I see the law. So lets all stand together and fight the inequities of our government, as well as their corruption and exploitation of our rights and civil liberties. Thank you for listening, and lets keep fighting the good fight!"

Yellow on the Outside, Shame on the Inside: Asian Culture Revealed

"Why do Asians really get straight A's? Why do Asians really become doctors and lawyers? Why do Asians really play the piano? Many people believe that the reason has to do with the pressure to perform and the pressure to conform, however, it goes much deeper than that—much, much deeper! This didactic novel reveals the truths about Asian culture, which will shock you to the marrow of your bones—and open a hidden world of long-guarded secrets".

Tom Cryer vs IRS: NOT GUILTY! "I was at the trial to hear the NOT guilty verdict come out. He was acquitted on 2 counts of willful failure to file and all tax evasion charges were dropped even before the trial started. This goes to show that the IRS is full of shit and that people are catching on about the fact that there's no law that requires the average American to pay an income tax.

This is how Tom won: He is a seasoned attorney of 30 years. He's even made laws. So if he can't find a law after 15 years of research and after reading the entire Internal Revenue Code (6,000 pages), then where the hell is the law?

Also, he's always said that the income tax is legal and it is constitutional BUT to whom does it apply to? Only to those involved in international trade and foreign commerce like corporations (61% of corporations, however, don't pay any income tax according to the GAO since they write off your labor as an expense and put the burden on your shoulders to pay it for them. And to think that your company loves you!)

It was an unanimous verdict of 12-0 NOT guilty. The jury isn't stupid. Neither are you. Demand that the IRS show you the law (hint: there's none)!

I made a guest appearance on a radio show in Austin: (July 14, 1:00PM) Please listen to what went on at the trial and what you can do to fight the complicit income tax! Thanks!?

One of these posts was posted by an individual who proceeded to attempt to bomb a local natural gas facility in support of the Sovereign Citizens movement (Post 2). The second post is by an individual was arrested for making threats against the president (Post 1).

While the traditional view is that those who engage in acts of terrorism are covert, recent work by Gill and Corner (2013), as well as others (e.g., Meloy and O'Toole 2011) has found that offenders often exhibit "leakage" in that they advertise their intent and/or grievance to others. Gill (2015b) referred to this as the "lone-actor dilemma; in that acts of terrorism are about messaging and hence they need to communicate their *purpose* in order for their action to have meaning. Specifically, in his study of 119 lone-actor terrorists Gill and Horgan (2013) identified that in 58.8% of cases the offender produced letters or public statements prior to the event outlining his/her beliefs (but not necessarily his/her violent intent).

What this means then is that in a large number of cases there are verbal, and public, expressions of ideology which could provide a critical source of insight into the degree of threat that a person holds. Research on the diagnostic value of language has surged as computerized text analysis tools have become more accessible. Within the last decade alone there have been hundreds of studies linking the language used by an individual to a wide range of psychological correlates (Tausczik and Pennebaker 2009). It has long been established that word use by an individual is a stable predictor of individual differences with self-reports of personality (Pennebaker and King 1999). One clear example of this is neuroticism,[3] in that individuals who score higher on neuroticism are more likely to use first-person singulars (e.g., I, me, mine) more frequently. Extraverts on the other hand use more positive emotion words (e.g., great, happy, amazing). Many of these earliest findings have since been replicated across multiple studies (e.g., Yarkoni 2010). This form of linguistic analysis has also extended to social media use, where linguistic analysis of tweets has been used to diagnose personality (Sumner et al. 2012), as well as personal values (Chen et al. 2014).

While very much in its infancy, these methods have begun to be applied to threat assessment. For example, an open-source project currently being completed by the authors used linguistic analysis to detect the differences in narratives used by those who post extremist sentiments online and engage in violent behavior (true hits), from those who pose extremist sentiments online and do not engage in violent behavior (false positives). This linguistic analysis takes two forms; the first uses narrative and discourse analysis. Narrative and discourse analysis can take many forms, but one form of analysis that may be applied to narratives of personal experience as well as to any other forms of discourse (e.g., expositions in textbooks, diatribes) is called dependency analysis and was first developed by Deese (1983). This analysis examines many aspects of discourse, including departure from the timeline of an event, aspects upon which a narrator elaborates, and what a narrator leaves unsaid but is logically required to make a discourse sensible to listeners. Dependency

---

[3] As described by *Psychology Today*, "Neuroticism, one of the Big 5 personality traits, is typically defined as a tendency towards anxiety, self-doubt, depression, shyness, and other similar negative feelings. Since all personality traits, including emotional instability, exist on a spectrum, everyone is a little neurotic to some degree—some people are just much more neurotic than others." Online at: https://www.psychologytoday.com/us/basics/neuroticism

analysis has been adapted for clinical use and tracks the following dimensions (McCabe and Bliss 2003):

1. Topic Maintenance: whether all utterances are on topic.
2. Event Sequencing: whether all events narrated are in chronological order or acknowledged as out of order by conjunctions.
3. Informativeness: whether all specific information necessary to understand an experience is provided or implied, or whether there are omissions of key information. A second aspect of informativeness assesses whether there is appropriate elaboration for the cultural context or whether the discourse seems truncated for the context, and the third aspect of informativeness necessary to assess is whether a) there is ample orientation as to the setting of an experience, b) there are enough events conveyed, and c) there is enough, too much, or appropriate evaluation of events (e.g., "those XXX are horrible people").
4. Referencing: whether or not we know what pronouns refer to, or whether they are ambiguous.
5. Conjunctive Cohesion: whether statements are connected by causal conjunctions (because, so), temporal conjunctions (e.g., then, before), simple coordinate conjunctions (and), or not connected by many conjunctions at all.

Dependency analysis has been used reliably and validated as one component of what nonprofessional listeners/readers factor into judging the quality of a narrative. For example, you can see that in the social media posts provided above, Post 1 maintains a far closer examination of the topic than Post 2.

A second form of analysis being conducted on this data is based on machine learning. Given the expansion of linguistic content available on almost any issues, sentiment analysis has become a mature machine learning research topic (see Pang and Lee 2008). Machine learning has been applied to many different domains, from movie reviews (Pang et al. 2002) to congressional floor debates (Thomas et al. 2006), product recommendations (Snyder and Barzilay 2007), and social media materials (e.g., Gautam and Yadav 2014). Currently, linguistic analysis has been applied to understanding the language of extremist groups (e.g., Prentice et al. 2012), as well as how to classify material as extremist or not (Wanner 2019). But it has not, to date, been used at the individual level. In this sense, machine learning can also be used to understand the (potential) differences in narratives and discourse and the degree to which those demonstrate intent to engage in violent extremist behavior. While the application of linguistic analysis to the areas of extremism is in its infancy, it does hold significant potential in supporting on-going decisions of risk.

6 Conclusion

To sum up, recent decades have seen an increasing use of technology for risk assessment and monitoring for potential terrorist activity. In particular, technologies have provided new opportunities to gather and analyze massive amounts of data from an increasingly broad spectrum of sources. Simulations, databases and advanced algorithms are being used by law enforcement and other government agencies to inform counterterrorism efforts. And yet, there is still much work to be done. This chapter has focused particular attention on the need to address low-base rates, problematic signal-to-noise ratios, and false flags that stem from the nature of Internet activities—all these continue to pose problems for any technological solution applied to countering terrorism. Meanwhile, some new and emerging technologies—like machine learning—show promise for improving the effectiveness of counterterrorism. With the right combination of investment in research and development (both in the private sector and by government agencies), and advanced analytical training that draws from psychology and other behavioral sciences, we remain confident that in the future technology can indeed help practitioners counter terrorism with increasing sophistication and success.

Bibliography

Bakker, E. (2006). *Jihadi terrorists in Europe, their characteristics and the circumstances in which they joined the Jihad.* Clingendael Security Paper: Netherlands Institute of International Relations.

BBC News. (2001, May 29). What you need to know about Echelon, *BBC News*. Online at: http://news.bbc.co.uk/2/hi/sci/tech/1357513.stm.

Berger, J. M., & Morgan, J. (2015). The ISIS twitter census: Defining and describing the population of ISIS supporters on twitter, *Brooking project on U.S. Relations with the Islamic World, Analysis Paper, 20.*

Bjelopera, J. P. (2017, August 21). *Domestic Terrorism: An Overview* (pp. 3). Washington, DC: Congressional Research Service. Online at: https://fas.org/sgp/crs/terror/R44921.pdf.

Bloom, R. M., & Clark, W. T. (2016, Spring). Small Cells, Big Problems: the increasing precision of cell site location information and the need for fourth amendment protections. *Journal of Criminal Law and Criminology* 106(2). Online at: https://scholarlycommons.law.northwestern.edu/cgi/viewcontent.cgi?article=7583&context=jclc.

Brachman, J. (2009). *Global Jihadism: Theory and practice* (p. 2009). London: Routledge.

Chen, J., Hseih, G., Mahmud, J., & Nichols, J. (2014). Understanding Individuals' Personal Values from Social Media Word Use. *Values & Social Norms,* CSCW 14, February 15–19.

Corner, E., & Gill, P. (2015). A false dichotomy? Mental illness and lone-actor terrorism. *Law and Human Behaviour, 39*, 23–34.

Deese, J. (1983). *Thought into speech: Psychology of a language*. Englewood Cliffs: Prentice-Hall.

Department of Justice Policy Guidance. (n.d.). *Domestic use of unmanned aircraft systems* https://www.justice.gov/file/441266/download;

Docobo, J. M. (2006). Protecting America's communities from terrorism: A law enforcement perspective. In J. J. F. Forest (Ed.), *Homeland security: Protecting America's targets*. Westport, CT: Praeger.

Dubrovsky, V. J., Kiesler, S., & Sethna, B. N. (1991). The equalization phenomenon: Status effects in computer- mediated and face-to-face decision-making groups. *Human-Computer Interaction, 6*, 119–146.

FBIs Counterterrorism and Forensic Science Research Unit (CFSRU), online at: http://www.fbi.gov/about-us/lab/scientific-analysis/counterterrorism-forensic-science-research.

Federal Bureau of Investigation. (2004, August 9). *Strategic plan 2002–2009*. Online at: http://www.fbi.gov/publications/strategicplan/stategicplantext.htm

Fernandes, R., Hatfield, J., & Soames Job, R. F. (2010). A systematic investigation of the differential predictors for speeding, drink-driving, driving while fatigued, and not wearing a seat belt, among young drivers. *Transportation Research Part F: Traffic Psychology and Behaviour, 13*(3), 179–196.

Forest, J. J. F. (2015). Introduction. In *Essentials of counterterrorism* (pp. 1–32). Westport: Praeger.
Friedersdorf, C. (2016, March 10). The rapid rise of federal surveillance drones over America. *The Atlantic*. Online at: https://www.theatlantic.com/politics/archive/2016/03/the-rapid-rise-of-federal-surveillance-drones-over-america/473136/.
Gallagher, R. (2015, November 18). From Paris to Boston, terrorists were already known to authorities. *The Intercept*. Retrieved via: https://theintercept.com/2015/11/18/terrorists-were-already-known-to-authorities/
Gautam, G., & Yadav, D. (2014). Sentiment analysis of twitter data using machine learning approaches and -semantic analysis. *2014 Seventh International Conference on Contemporary Computing* (IC3), Noida, 2014, pp. 437–442.
Gelder, K. (2006). Epic fantasy and global terrorism. In E. Mathijs & M. Pomerance (Eds.), *From hobbits to Hollywood: Essays on Peter Jackson's Lord of the rings* (pp. 101–118). Kenilworth: Rodopi.
Gill, P. (2013). Tipping point-the adoption of suicide bombings. *Psicología Política, 46*, 77–94.
Gill, P. (2015a). Toward a scientific approach to identifying and understanding indicators of radicalization and terrorist intent: Eight key problems. *Educational Publishing Foundation, 2*(3–4), 187.
Gill, P. (2015b). *Lone actor terrorism: A Behavioral analysis*. London: Routledge.
Gill, P., & Corner, E. (2013). Disaggregating terrorist offenders: Implications for research and practice. *Criminology & Public Policy, 12*(1), 93–101.
Gill, P., & Horgan, J. (2013). Who were the volunteers? The shifting sociological and operational profile of 1240 provisional Irish Republican Army members. *Terrorism and Political Violence, 25*(4), 435–456.
Gill, P., Horgan, J., & Deckert, P. (2014). Bombing alone: Tracing the motivations and antecedent behaviors of lone-actor terrorists. *Journal of Forensic Sciences, 59*(2), 425–435.
Gruenewald, J., Chermak, S., & Freilich, J. D. (2013a). Far-right lone wolf homicides in the United States. *Studies in Conflict and Terrorism, 36*, 1005–1024.
Gruenewald, J., Chermak, S., & Freilich, J. D. (2013b). Distinguishing 'loner' attacks from other domestic extremist violence: A comparison of far-right homicide incident and offender characteristics. *Criminology & Public Policy, 12*(1), 63–64.
Hoffman, B. (2006). *Inside Terrorism* (pp. 40–41). New York: Columbia University Press.
Horgan, J. (2013). *The psychology of terrorism* (2nd ed.). New York: Routledge.
Horgan, J., Shortland, N., Abbascianno, S., & Walsh, S. (2016). Actions speak louder than words: A behavioral analysis of 183 individual convicted of Terrorist Offenses in the United States from 1989–2012. *Submitted to Journal of Forensic Sciences.*
Horgan, J., Shortland, N. D., & Abbasciano, S. (2018). Towards a typology of terrorism involvement: A behavioral differentiation of violent extremist offenders. *Journal of Threat Assessment and Management, 5*(2), 84–102.
Intelligence and Security Committee of Parliament. (2014). *Report on the intelligence relating to the murder of Fusilier Lee Rigby.* United Kingdom. Online at: http://isc.independent.gov.uk/committee-reports/special-reports
International Association of Chiefs of Police. (2005, May 17). *From hometown security to homeland security: IACP's principles for a locally designed and nationally coordinated homeland security strategy*. Online at http://www.theiacp.org/leg_policy/HomelandSecurityWP.PDF.
Kebbell, M. R., & Porter, L. (2011). An intelligence assessment framework for identifying individuals at risk of committing acts of violent extremism against the west. *Security Journal, 25*, 212–228.
Kelly, E. V. (2018, January 11). What is the Section 702 surveillance program and why should you care? *USA Today*. Online at: https://www.usatoday.com/story/news/politics/2018/01/11/what-section-702-surveillance-program-and-why-should-you-care/1025582001/.
Khader, M., Neo, L. S., Ong, G., Tan, E., & Chin, J. (Eds.). (2016). *Countering violent extremism and radicalisation in the digital era*. Hershey: IGI Global.
Kraft, M. (2015). The US Governments counterterrorism research and development programs. In J. J. F. Forest (Ed.), *Essentials of counterterrorism* (pp. 195–220). Westport: Praeger.

Kraft, M., & Marks, E. (2012). *U.S. Government counterterrorism: A guide to who does what*. Washington, DC: CTC Press.

Lankford, A. (2013). *The myth of martyrdom: What really drives suicide bombers, rampage shooters, and other self-destructive killers*. New York: Palgrave Macmillan.

Lum, C., Kennedy, L. W., et al. (2006) Are Counter-Terrorism Strategies Effective? The Results of the Campbell Systematic Review on Counter-Terrorism Evaluation Research, *Journal of Experimental Criminology* : 489–516.

Matney, L. (2015). *Uncovering ECHELON: The Top-Secret NSA/GCHQ Program That Has Been Watching You Your Entire Life*. TechCrunch (August 23). Online at: https://techcrunch.com/2015/08/03/uncovering-echelon-the-top-secret-nsa-program-that-has-been-watching-you-your-entire-life/.

McCabe, A., & Bliss, L. S. (2003). Personal narratives: Assessment and intervention. *Perspectives on Language Learning and Education, 19*(4), 130.

Meloy, J. R. (2011). Approaching and attacking public figures: A contemporary analysis of communications and behavior. In C. Chauvin (Ed.), *Threatening communications and behavior: Perspectives on the pursuit of public figures* (pp. 75–101). Washington, DC: National Academies Press.

Meloy, J. R., & Gill, P. (2016). The lone-actor terrorist and the TRAP-18. *Journal of Threat Assessment and Management, 3*(1), 37–52.

Meloy, J. R., & O'Toole, M. E. (2011). The concept of leakage in threat assessment. *Behavioral Sciences & the Law, 29*, 513–527. https://doi.org/10.1002/bsl.986.

Meloy, J. R., & Yakeley, J. (2014). The violent true believer as a "lone wolf"—Psychoanalytic perspectives on terrorism. *Behavioral Sciences & the Law, 32*, 347–365.

Office of the Director of National Intelligence. (2018). *Statistical transparency report, calendar year 2017*. Washington, DC: Office of Civil Liberties, Privacy and Transparency (May 4). Online at: https://www.dni.gov/files/documents/icotr/2018-ASTR%2D%2D%2D%2DCY2017%2D%2D%2D%2DFINAL-for-Release-5.4.18.pdf.

Pang, B., & Lee, L. (2008). Opinion mining and sentiment analysis. *Foundations and Trends in Information Retrieval, 2*(1–2), 1–135.

Pang, B., Lee, L., & Vaithyanathan, S. (2002). Thumbs up? Sentiment classification using machine learning techniques. In *Proceedings of the Conference on Empirical Methods in Natural Language Processing*.

Pennebaker, J. W., & King, L. A. (1999). Linguistic styles: Language use as an individual difference. *Journal of Personality and Social Psychology, 77*, 1296–1312.

Postmes, T., Spears, R., & Lea, M. (2002). Intergroup differentiation in computermediated communication: Effects of depersonalization. *Group Dynamics, 6*, 3–16.

Prentice, S., Rayson, P., & Taylor, P. J. (2012). The language of Islamic extremism: Towards an automated identification of beliefs, motivations and justifications. *International Journal of Corpus Linguistics, 17*(2), 259–286.

Prentice-Dunn, S., & Rogers, R. W. (1989). Deindividuation and the self-regulation of behavior. In P. B. Paulus (Ed.), *The psychology of group influence* (2nd ed., pp. 86–109). Hillsdale: Lawrence Erlbaum.

Pressman, D. E., & Flockton, J. (2014). Violent extremist risk assessment: Issues and application of the VERA-2 in a high–security correctional setting. In A. Silke (Ed.), *Prisons, terrorism and extremism: Critical issues in management, radicalisation and reform* (pp. 122–143). New York: Routledge.

Pressman, E. D., & Ivan, C. (2016). Internet use and violent extremism: A cyber-VERA risk assessment protocol. In M. Khader, L. S. Neo, G. Ong, E. Tan, & J. Chin (Eds.), *Countering violent extremism and radicalisation in the digital era* (pp. 391–409). Hershey: IGI Global.

Risen, J., & Lichtblau, E. (2005, December 16). Bush Lets U.S. Spy on Callers Without Courts. *The New York Times*. Online at: https://www.nytimes.com/2005/12/16/politics/bush-lets-us-spy-on-callers-without-courts.html

Roberts, K., & Horgan, J. (2008). Risk assessment and the terrorist. *Perspectives on Terrorism, 2*(6), 1–4.

Rosenau, W. (2014). *The 'first war on terrorism?' – U.S. Domestic counterterrorism during the 1970s and early 1980s*. Washington, DC: CNA Center for Strategic Studies (October).

Sageman, M. (2004). *Understanding terror networks*. Philadelphia: University of Philadelphia Press.

Sageman, M. (2014). The stagnation in terrorism research. *Terrorism and Political Violence, 26*, 565–580.

Sarma, K. M. (2017). Risk assessment and the prevention of radicalization from nonviolence into terrorism. *American Psychologist, 72*(3), 278–288.

Schmid, A. (2013). *The Routledge handbook of terrorism research*. New York: Taylor & Francis.

Schmid, A. P. (2014). Comments on Marc Sageman's polemic "the stagnation in terrorism research". *Terrorism and Political Violence, 26*(4), 587–595.

Schuurman, B. (2018). Research on terrorism, 2007–2016: A review of data, methods, and authorship. *Terrorism and Political Violence*, 1–16.

Silke, A. (2007). The impact of 9/11 on research on terrorism. In *Mapping terrorism research* (pp. 90–107). New York: Routledge.

Simcox, R., & Dyer, E. (2013). *Al Qaeda in the United States: A complete analysis of terrorism offenses*. London: The Henry Jackson Society.

Snyder, B., & Barzilay, R. (2007). Multiple aspect ranking using the good grief algorithm. in *Proceedings of the Conference of the North American Chapter of the Association for Computational Linguistics: Human Language Technologies (NAACL/HLT-2007)*.

Stern, J. (2014). Response to Marc Sageman's "the stagnation in terrorism research". *Terrorism and Political Violence, 26*(4), 607–613.

Suler, J. (2004). The online disinhibition effect. *Cyber Psychology and Behavior, 7*, 321–326.

Sumner, C., Byers, A., Boochever, R., & Park, G. J. (2012). Predicting dark triad personality traits from Twitter usage and a linguistic analysis of tweets. *11th International Conference on Machine Learning and Applications (ICMLA)*, vol. 2. IEEE Press.

Talbot, D. (2005). Terror's server. *Technology Review*. Retrieved from http://www.technologyreview.com/infotech/14150/.

Tausczik, Y. R., & Pennebaker, J. W. (2009). The psychological meaning of words: LIWC and computerized text analysis methods. *Journal of Language and Social Psychology, 29*(1), 24–54.

Taylor, M. (2014). If I were you, I wouldn't start from here: Response to Marc Sageman's "the stagnation in terrorism research". *Terrorism and Political Violence, 26*(4), 581–586.

Thomas, M., Pang, B., & Lee, L. (2006). Get out the vote: Determining support or opposition from congressional floor-debate transcripts. *Proceedings of the Conference on Empirical Methods in Natural Language Processing (EMNLP)*, pp. 327–335.

Wanner, L. (2019). Automatic classification and linguistic analysis of extremist online material. *International Conference on Multimedia Modeling*.

Webster Commission. (2012). *Final Report of the William H. Webster Commission on Federal Bureau of Investigation, Counterterrorism Intelligence, and the Events at Ford Hood, Texas, on November 5, 2009*. Online at: https://www.hsdl.org/?view&did=717443

Weimann, G. (2004). www.terror.net: How modern terrorism uses the internet. *Special Report: United States Institute of Peace, 116*, 1–12.

Weimann, G. (2006). *Terror on the internet: The new arena, the new challenges*. Washington, DC: United States Institute of Peace.

Yarkoni, T. (2010). Personality in 100,000 words: A large-scale analysis of personality and word use among bloggers. *Journal of Research in Personality, 44*, 363–373.

Zimbardo, P. G. (1969). The human choice: Individuation, reason, and order versus deindividuation, impulse, and chaos. *Nebraska Symposium on Motivation, 17*, 237–307.

Human Trafficking and the Darknet: Technology, Innovation, and Evolving Criminal Justice Strategies

Joan Reid and Bryanna Fox

Abstract In an era of unprecedented technological accessibility coupled with the possibility of online anonymity, human traffickers often operate in the shadows – avoiding interference by typical societal safeguards and law enforcement. This chapter highlights challenges to combatting human trafficking due to the widespread use of mobile technology and the anonymity of the Darknet. The chapter also reviews criminal justice resources devoted to combatting human trafficking facilitated by the Internet.

Keywords Human trafficking · Mobile technology · Darknet · Criminal justice

1 Introduction

Human trafficking, also referred to as trafficking in persons and modern-day slavery, has long been considered a human rights issue that deprives millions of men, women, and children of their fundamental rights to human dignity and personal freedoms. Due to concerted efforts by international and national communities, human trafficking is now considered a complex problem to be confronted and thwarted by the criminal justice systems around the world. As a criminal enterprise, trafficking in persons has expanded into a global problem which fuels corruption and violence, finances gangs and organized crime, and threatens the safety and security of hundreds of thousands of people.

J. Reid
Department of Criminology, University of South Florida-St. Petersburg,
St. Petersburg, FL, USA
e-mail: jareid2@usf.edu

B. Fox (✉)
Department of Criminology, University of South Florida, Tampa, FL, USA
e-mail: bhfox@usf.edu

B. Fox et al. (eds.), *Science Informed Policing*, Advanced Sciences and Technologies for Security Applications,
https://doi.org/10.1007/978-3-030-41287-6_5

Trafficking in persons takes many forms, and exists in nearly every conceivable job setting or industry. Human trafficking can be found in the sex trade, both in prostitution and pornography, in restaurants or bars, in street gangs or the drug trade, in service industries such as tourism or health care, in domestic servitude, in manufacturing textiles or food processing, in agriculture or construction, in a diamond mine, and in a despot's army. The fragmented nature of human trafficking frustrates cooperation between law enforcement agencies, and impedes victim identification and recovery.

Complicating the issue further, in the last decade human trafficking has gone digital and dark. Globalization and technology facilitate the spread of human trafficking by multiplying connections and opportunities for traffickers to recruit and exploit victims. In an era of unprecedented technological accessibility, coupled with the perception of almost complete online anonymity, concerns have arisen regarding the escalation and detectability of human trafficking in its most egregious forms. For example, child pornographic exploitation, a form of sex trafficking, thrives on the dark web. A study of services on the dark web found that 83% of hidden service requests were for online destinations delivering child pornographic content, dwarfing the number of requests for any of the other categories of content type such as illegal drugs, gambling, bitcoin-related sites, or anonymous whistle-blowing (Owen 2014). Former U.S. Attorney General Eric Holder (2011) stated that in the last decade there has been "a historic rise in the distribution of child pornography, in the number of images being shared online, and in the level of violence associated with child exploitation and sexual abuse crimes. Tragically, the only place we've seen a decrease is in the age of victims. This is – quite simply – unacceptable."

Here's how this chapter is organized. Section 2 includes an analysis of the evidence pertaining to the extent and nature of human trafficking, with emphasis of how advances in technology are facilitating this crime. Section 3 provides an overview of the dark web and how human traffickers are using the dark web to avoid detection by law enforcement. Section 4 examines criminal justice strategies and resources to combat human trafficking facilitated by digital technology and the Internet.

2 Human Trafficking Goes Digital

The illegal business of trafficking in persons involves three components: (1) the grooming and recruitment of vulnerable victims, (2) the advertisement of illegal services provided by human trafficking victims to a potential client-base, and (3) the payment for business expenses and services. Consequently, human trafficking is largely facilitated by digital technologies such as mobile phones and social networking sites (Latonero 2011). For instance, web-enabled mobile phones facilitate these activities while offering some assurance of anonymity for both traffickers and consumers. Countless examples of human trafficking cases involve the grooming and recruitment of vulnerable victims via Myspace, Facebook, online websites of traffickers, or other social networking sites (Latonero 2011; Reid 2016a). Advertising of services provided by human trafficking victims to potential clients have been observed on social networking and online classified advertising sites such as Craigslist, Backpage, and Facebook (Latonero 2011). Human traffickers and consumers also take advantage of the simplicity provided by the Internet and credit cards to pay for business expenses or services – to secure a hotel room or reserve many rooms in various cities, to book online ads on adult sites, to make money transfers, to purchase frequent fast-food deliveries, book a ride-sharing service, or to purchase airline tickets.

2.1 The Use of Technology in Sex Trafficking

Digital and online technology has impacted all forms of sex trafficking and prompted the development of several emergent forms (Reid 2016b). An explosion in the child pornographic material, documented as both a quantitative upsurge in the number of producers and consumers as well as substantial qualitative degeneracy of child pornographic material, has been attributed to the unprecedented technological accessibility and greater assurance of anonymity available through the Internet. Today, images, recordings, and even live performances involving the sexual exploitation of children can be assessed through various types of Internet technology including websites, email, instant messaging, P2P (peer-to-peer or person-to-person) file sharing networks, and social networking sites (Beech et al. 2008; Quayle and Jones 2011; Wortley and Smallbone 2012; Wolak et al. 2014). Consumers of child pornographic material are able to connect to live images or real-time video productions of children being sexually abused through the availability of video conferencing and Webcams. Viewers of live sexual abuse of children can even make requests in real time and buy instant gratification of specific sexual fetishes as the abuse is being broadcast – a new trend labeled "molestation-on-demand" (Beech et al. 2008; Gonzales 2006).

One emerging example of the impact of globalism and the Internet, *webcam child sex tourism*, involves wealthy consumers of child sexual exploitation, frequently from the United States or Western Europe, sexually exploiting disadvantaged children living in other countries. Using the availability of online payment, these globalized offenders direct and view children in other countries to perform sexual acts in front of live streaming webcams (Terres des hommes 2013). Parents or child sex ring organizers transmit live sexual activity or abuse of children while the customers demand specific requests (Rodas 2014). The Peoples Recovery Empowerment Development Assistance (PREDA) Foundation, a social service agency in the Philippines, estimates that tens of thousands of Filipino children are sexually abused annually through webcam child sex tourism (Terres des hommes 2013).

Online businesses, such as modeling agencies, operate on the Internet and on the edge of legality to recruit and exploit victims (U.S. Department of Justice [DOJ]

2013; Reid 2008; "Florida No Haven for Teen Exploitation," 2008). These modeling agencies, described as displaying young girls "on the Internet wearing nothing more than bras and G-strings and connecting them to pedophiles in online chats", were eventually shut down in the State of Florida through the passage of more stringent state legislation ("Florida No Haven for Teen Exploitation," 2008, para. 6). Producers of teen modeling images and videos stay abreast of changes in state statutes that regulate what is legal and illegal, and take advantage of ambiguities in state statutes to avoid arrests and prosecution. Moreover, human traffickers on social media frequently take on the guise of model agency scouts seeking "webcam models" in order to recruit and entrap young teens in sex trafficking (Latonero 2011).

While child pornographic exploitation has been transformed by digital technology, other forms of sex trafficking have also benefited from the technological advances. Sex trafficking involving forced prostitution or prostitution of minors largely occurs online. Sex traffickers use smart phones and computers to facilitate sex trafficking in various ways – for recruiting victims, advertising victims, to communicate with other offenders (Armagh et al. 2006; Beech et al. 2008; Finn and Stalans 2016; Latonero et al. 2012; Reid 2015, 2018). The use of the Internet to sexually exploit others has escalated in recent years; it has been estimated that 75% of sex trafficking of minors in the United States is facilitated using the Internet, in comparison to approximately 38% prior to 2004 (Bouche and THORN 2015). In a survey of 115 sex trafficking survivors, Bouche and THORN (2015) found that 63% reported being advertised online, with almost half reporting that they were advertised on Backpage, Craigslist, or Facebook. A recent investigative report on sex trafficking in the U.S. found that "The women are often sold online – and raped in person. Much of the police work is done on the Internet, a task that can weigh on investigators enduring long hours and reviewing endless streams of disturbing content" (Keiper and Chiaramonte 2019a). This combination of increased availability of marketing in the ubiquity of the Internet, associated anonymity, and challenges these pose to law enforcement has created a perfect storm for sex trafficking to proliferate on the Internet in front of our eyes. As Pasco Sheriff's Office Corporal and Chair of the Pasco County Human Trafficking Task Force Alan Wilkett stated about sex traffickers on the Internet: "They're hiding in plain sight" (Keiper and Chiaramonte 2019a).

The Internet has impacted other less recognized forms of human trafficking. For example, mail order brides are no longer recruited and ordered via the mail, as today they are recruited and advertised on online dating sites or International marriage agencies and more fittingly called "Internet brides" (Jones 2011). Prior to the advent of the Internet, most mail-order brides were brought to market through letters, newspaper advertisements, and portraits. When using the international postal system, connecting the "bride" with a buyer took a long time. Today, the Internet, with its capacity to reach brides and buyers all over the world anonymously, has become one of the most common vehicles for shopping for a spouse. While many engage in this exchange and marriage willingly, others – many young girls – do not. Currently, there is an ongoing investigation into child bride immigration into the United States. Between 2007 and 2017, U.S. Citizenship and Immigration Services (USCIS) approved 8686 petitions for spousal or fiancé entry into the United States involving minors. "Two minors whose petitions were approved were 13 years old; 38 were 14 years old; 269 were 15 years old; 1,768 were 16 years old; and the remaining 6,609 were 17 years old" (Johnson 2019, p. 2). The opportunity to sponsor a visa for a child spouse or fiancé in the United States has been deemed a "federal loophole" which leaves "the most vulnerable children, often girls promised to adult men, defenseless to families who hope to offer U.S. citizenship in exchange for a marital agreement." (Johnson 2019, p. 7).

2.2 *The Use of Technology in Labor Trafficking*

While research on the use of digital technology in sex trafficking has drawn the attention of researchers and policymakers, less research has been conducted on the use of digital technology in labor trafficking. One of the rare studies on this topic featured a 2015 case study based in the Philippines (Latonero et al. 2015). The researchers found that traditional, non-technological modes of recruitment for migrant workers – who are particularly vulnerable to labor trafficking – are still widely used such as newspapers, billboards, job fairs, and word-of-mouth. Nevertheless, Latonero and colleagues (2015) also found evidence of the budding use of online job sites, social media, email, and mobile phones in recruiting and advertisement of job opportunities for migrant workers. As with sex trafficking, these new opportunities for recruitment using digital technology are less expensive and more effective for employers seeking to hire workers than traditional methods. These online recruiting modalities were exposed as the primary way that online technology is being used in labor trafficking (Latonero et al. 2015).

A 2015 report by the United Nations Office on Drugs and Crime (UNODC) provides a list of red flags that could indicate that an online (or traditional) recruitment agency is engaging in labor trafficking. The list of red flags include, but are not limited to: workers required to sign more than one contract or a contract in a language unfamiliar to them; bank accounts for workers opened under the name of a staff member of the recruitment agency; recruitment agencies are unregistered/not licensed or not licensed in the workers' home country or not licensed for the work sector for which they are recruiting; tourist agencies that recruit workers; recruitment agencies involved with moneylenders in the country of origin or recruitment agencies in the destination country. Importantly, one common method that is used by exploitive recruitment agencies to entrap workers in labor trafficking is charging a "recruitment fee" which creates indebtedness for the workers (UNODC 2015a).

> Recruiters and recruitment agencies often promise workers high salaries and excellent conditions of work and charge workers large recruitment fees – sometimes legal, sometimes illegal. Fees can cover a variety of costs, including travel, passport and visa processing, medical exams, and unspecified fees and service charges, much of which may go directly to the recruiter (UNODC 2015a, p. 8).

For example, for over a decade one globally operating recruitment agency fraudulently recruited and exploited hundreds of migrant workers from Latin America, South East Asia, Eastern Europe, Central Asia, and elsewhere, to work in the United States through false promises of good employment. The workers were promised high-paying jobs, free housing, transportation, food, and other provisions as part of the recruitment package. The recruitment agency charged workers recruitment fees up to $5,000. Workers were also charged visa renewal fees of $3,000 and families were threatened with additional fees. When the workers arrived, they were forced to live in very overcrowded apartments with overpriced rents and coerced to work through the use of threats of deportation and by the withholding of wages (UNODC 2015a).

2.3 The Use of Technology in Other Forms of Human Trafficking

While the majority of human trafficking cases can be categorized as either sex or labor trafficking, it is important to note emerging forms of human trafficking and the role of technology in their emergence. Other less researched forms of human trafficking organ trafficking and human trafficking for the purpose of organ harvesting (Caulfield et al. 2016). Human trafficking for the purpose of organ removal differs from trafficking in organs. With trafficking in organs, the organ is removed and trafficked from the donor location to the recipient location. In human trafficking for organ harvesting, individuals are trafficked to the location of the organ recipients and their organs are removed on location. This is sometimes referred to as "transplantation tourism" and involves multiple levels of organization in the destination country. Conservative estimates of the annual profits of illicit organ trade range between $840 million to $1.7 billion (Global Financial Integrity 2017). Organ trafficking and trafficking persons for the purpose of organ harvesting involve a complex collection of actors including organ recipients (sometime complicit and sometimes unaware of illegality of the transplant), medical professionals (surgeons, nurses, kidney specialists), recruiters, brokers, laboratory technicians, medical facility operators and staff, insurance companies, travel agencies, airlines and their staff, as well as guards, drivers, service providers, law enforcement officials, and translators (UNODC 2015b).

The Palermo Protocol of 2000[1] provides the foundational definition for human trafficking that is most commonly used as a blueprint for national anti-human trafficking laws. The Palermo Protocol specifically includes organ trafficking as a form of human trafficking exploitation.

[1] Protocol to Prevent, Suppress and Punish Trafficking in Persons. Especially Women and Children, Supplementing the United Nations Convention Against Transnational Organized Crime, U.N. Doc. A/53/383 (2000) [Palermo Protocol], available at http://www.uncjin.org/Documents/Conventions/dcatoc/final_documents_2/convention_%20traff_eng.pdf

> "Trafficking in persons" shall mean the recruitment, transportation, transfer, harbouring or receipt of persons, by means of the threat or use of force or other forms of coercion, of abduction, of fraud, of deception, of the abuse of power or of a position of vulnerability or of the giving or receiving of payments or benefits to achieve the consent of a person having control over another person, for the purpose of exploitation. Exploitation shall include, at a minimum, the exploitation of the prostitution of others or other forms of sexual exploitation, forced labour or services, slavery or practices similar to slavery, servitude or the removal of organs (Article 3, paragraph a).

In the United States, neither organ trafficking nor human trafficking for organ harvesting are included in the federal anti-human trafficking law (Pugliese 2007; Trafficking Victims Protection Act [TVPA] of 2000). While not included in the TVPA, U.S. federal law prohibits the purchase or sale of organs (42 U.S.C. sections 273–274[2]). Most U.S. States follow the federal definition of human trafficking and do not include organ trafficking in anti-trafficking laws. Minnesota and Massachusetts are exceptions, as these states include organ trafficking in their anti-human trafficking statutes. While not uniformly included in anti-human trafficking laws, the illegality and unethical practices surrounding transnational organ transplantation is clear (Delmonico 2009; Pugliese 2007; Scheper-Hughes 2003). For example, during 2006, it was estimated that at least 4,000 prisoners were executed in China to provide approximately 8,000 kidneys and 3,000 livers for patients, the majority of whom were foreigners, purchasing these organs (Delmonico 2009). "In general, the [transnational] circulation of kidneys follows the established routes of capital from South to North, from poorer to more affluent bodies, from black and brown bodies to white ones, and from females to males, or from poor males to more affluent males. Women are rarely the recipients of purchased or purloined organs anywhere in the world." (Scheper-Hughes 2003, pp. 199–200).

Trafficking in organs and trafficking in persons for the purpose of organ harvesting are facilitated by "international Internet brokers [who] prey on consumer prejudices and on the anxieties of transplant candidates. For example, "'Livers-4-You', until a few weeks ago when it was forced off cyberspace, advertised the following via kidneysurgery@s-s-net.com: 'Want a healthy living donor next week, or a morgue organ in 5 years? We are a new organization with a New York City phone number and unique experience in locating the overseas pathway for those waiting too long for a transplant.'" (Scheper-Hughes 2003, p. 214). Similarly, a victim of organ trafficking reported that he was browsing the Internet desperately looking for ways to pay for his college bills and for his father's medical treatment when he found a Russian medical message board that stated: "Become a donor of a kidney for some money!" (UNODC 2015b, p. 44). The young man sent an email to the email address provided online, and the next day received an email response offering him EUR 10,000 for a kidney (UNODC 2015b). Following the transplant and his return to his home country, the young man was contacted via phone by the organ transplant recruiter and offered $1,000 for referrals. The recruiter also threatened him to not go to the police, stating "'if you go to the police you can disappear"

[2] Organ procurement organizations https://www.law.cornell.edu/uscode/text/42/273

(UNODC 2015b, p. 44). Moreover, transnational organ transplantation is often advertised by governmental tourism ministries websites. In India, "the Ministry of Tourism's *Incredible India* website advertises kidney transplants and notes the comparatively 'cheap' price of renal transplants at Indian medical facilities" (Turner 2008, p. 391). Hence, while most countries suffer from organ shortages, governmental tourism agencies continue to advertise the availability of organs to foreigners in desperate need of organ transplants to bolster their countries' gross national product.

Additional booming industries that can involve human trafficking and exploitive practices are trafficking in babies for adoption, and trafficking in persons for the purpose of birthing babies or "baby harvesting". Baby harvesting occurs in "baby factories" where young women are held captive and used like industrial vaginas to produce babies for sell (Nwaka and Odoemene 2019). Similar to organ trafficking, many levels of corruption and collusion with medical professionals are required to carry out this type of human trafficking. For example, 14 people were recently arrested in Peru, including the former director of Peru's National Police, as part of an operation to shut down a baby trafficking ring. Pregnant women, the parents of a 5-month-old rescued baby, women acting as recruiters of pregnant women, a gynecologist, and a pediatrician were all arrested as part of the operation (Brennan 2018). At the time of the operation, police were investigating whether the children were being sold for adoption to foreign couples or whether they were being sold to organ trafficking rings (Brennan 2018). Trafficking in babies for adoption and baby harvesting is facilitated by the Internet. The Internet provides information and publicity and it crucial for connecting individual buyers and sellers (Gupta 2018).

2.4 Normalization of Human Trafficking Through Advances in Technology

One seldom discussed consequence of the Internet is the ease of connectivity between participants and facilitators of human trafficking, such as consumers, distributors, and producers of child pornographic content. Prior to the advent of the Internet, producing, reproducing, and anonymously distributing child pornographic material was very difficult. Consumers of child pornographic material were not able to easily find each other and purchasing or trading child pornographic images was extremely risky (Gonzales 2006). Before easy accessibility and anonymity was facilitated by the Internet, child pornographic offenders were perceived as socially isolated sex offenders. With the Internet came unprecedented license for individuals to create, access, share, and trade images of child pornographic images anywhere in the world with ease – at the click of a button (Beech et al. 2008; Klain et al. 2001; Wolak et al. 2011; Wortley and Smallbone 2006; Murphy et al. 2014). Online communities made up of child pornographic consumers and producers are now able to attract new members and urge them to become involved in the sexual exploitation of children. A BBC investigation found a number of secret groups on Facebook, "created by and run for men with a sexual interest in children, including one being administered by a convicted paedophile who was still on the sex offenders' register" (Crawford 2016, para. 4).

For instance, in June 2019, the U.S. Department of Homeland Security and U.S. Marshals Service arrested a man who was deported from the U.S. in 2003 after sexually abusing a child for re-entering the U.S. illegally and then posting and distributing child pornography on Facebook (U.S. Department of Homeland Security 2019). In cases such as this, trading in images of child sexual abuse is part of new members' initiation into this online community – new images of previously unseen children are the premium trading commodity or community currency (Quayle and Taylor 2002; Murphy et al. 2014). One result of easy access to online child pornographic material is an increase in child sexual abuse triggered by activating, validating, and normalizing sexual deviant sexual urges in viewers of child pornographic material that were previously suppressed. As validation of this phenomenon, research on hidden services available on the Darknet noted:

> Several communities geared towards discussing and sharing illegitimate fetishes were readily available, and appeared to be active .... Users exchanged experiences and preferences, and even traded content. One notable example from a website called Pedo List included a commenter freely stating that he would 'Trade child porn. Have pics of my daughter.' (Moore and Rid 2016, p. 23).

Similarly, online "john" boards are now available that facilitate the trading of information regarding buying sex with women and girls in various cities in the United States, and around the globe. Research on the content and use of these online forums indicate that they are used to create "a brotherhood of men who buy sex, ... evoke images of a camp, a training ground with sets of rules that are actively taught and reinforced, and an 'old boys' club,' as one john puts it. This 'club' inculcates men throughout Illinois in the etiquette and social organization of the commercial sex industry in their communities" (Janson et al. 2013, p. 19).

These online forums promote commercial sexual activity by elevating the status of those with the most experience buying sex and commodifying women and girls. Depictions of women and girls included objective and descriptive information of women sold or sex acts provided (e.g., "COF—Cum on Face," "BSW—Black Street Walker"), subjective numeric rating systems (based on "looks, attitudes and service", e.g., 2/9/8 = Unattractive, but very friendly and good sex. 10/5/4 = Model material, but a poor attitude, and mediocre sex"), and reviews (e.g., "ROB—Rip Off Bitch") (Janson et al. 2013). One abbreviation "TOFTT – Take One for the Team" was an abbreviation used to reference reviews given regarding new, non-reviewed women or girls (Janson et al. 2013) and illustrates the comradery, connectedness, and loyalty created among users of these online sites.

3 Human Trafficking Goes Dark

Beyond use of digital technology to facilitate human trafficking on what is typically referred to as the "surface" or "searchable" Internet, human traffickers and consumers of illegal services also are benefiting from the protocols specific to the so-called "Darknet" or "Dark Web" segment of the Internet (Aked 2011; Bartlett 2014). Protocols of the Darknet are purposefully designed to protect user anonymity, and therefore these protocols assist human traffickers to avoid detection by law enforcement. Several studies of the Darknet have documented that the widespread use of non-standard protocols, anonymous IP address allocation, and extensions of peer-to-peer content sharing are enabling the development of criminal activity. In particular, consumers, distributors, and producers of child pornographic material are taking advantage of these criminogenic features of the Darknet.

As mentioned, the Internet consists of two parts: (1) the Surface Web that is indexed and accessible to search engines, and (2) the Deep Web that is space on the Internet that is not discoverable and not indexed by search engines. One study estimated that the size of the Deep Web is more than 550 billion pages, whereas the size of the Surface Web is only 1% of the Deep Web (i.e., one billion pages) (Bergman 2001). Content on the Deep Web mostly consists of innocuous material such as private company documents, databases maintained by governmental departments, and library archives that can be accessed via a search a particular website. Similarly, the Darknet (or Dark Web) consists of a small portion of the Deep Web that is intentionally hidden, and not accessible through standard web browsers or search engines. The Darknet has become a haven for criminals and illicit activities because of the anonymity it provides.

Anonymity (or bypassing censorship) on the Internet can be accomplished by using TOR software. There are other similar applications or so-called *privacy enhancing technologies*; however, TOR is the most prominent and famous tool among the available internet privacy and anonymity solutions (Çalışkan et al. 2015). TOR was named "The Onion Router" because is built on the concept of onion routing or layering your online activity or traffic similar to the layers of an onion. The technology was originally developed by the U.S. Naval Research Lab and Defense Advanced Research Projects Agency (DARPA), an agency of the U.S. Department of Defense, to protect political dissidents in countries with tight control over the Internet and individuals concerned about personal privacy (Çalışkan et al. 2015). Today the TOR project is an open source project that provides free software and an open network of servers that shield users from any type of surveillance while online. The TOR Project is a U.S. 501(c)(3) non-profit organization advancing human rights and freedoms by creating and deploying free and open-source anonymity and privacy technologies, supporting their unrestricted availability and use, and furthering their scientific and popular understanding.[3] The mission of the TOR project is "To advance human rights and freedoms by creating and deploying free and open

[3] TOR, The TOR Project. https://www.torproject.org/

source anonymity and privacy technologies, supporting their unrestricted availability and use, and furthering their scientific and popular understanding."[4]

In brief, when using an onion router, such as TOR, the connection from source A (e.g. the computer of an individual attempting to buy an illegal service) to destination B (e.g. the website of a provider of the illegal service) takes a detour along an encrypted route or chain, which is called an onion (Çalışkan et al. 2015). This creates a private, untraceable network pathway or circuit. The purpose of the encrypted route is to hide the locations or IP addresses of users who are browsing the web. All network communication along the route within the onion is encrypted and each node, also known as a relay, is only privy to the information about the adjacent nodes (i.e. the immediate sender and the next recipient). In this way, the complete picture of the communication route or chain is concealed. This circuit of nodes is comprised of at least three components – an entry or guard node, a relay node, and an exit node. This network chain hides user traffic, so that others cannot determine what websites users are accessing. As the number of nodes increases, so does the complexity and number of encrypted channels – resulting in greater assurance of anonymity. TOR is not the only available onion router. Others are available, such as I2P[5] and Freenet.[6]

The second purpose of TOR is to hide the locations of hundreds of websites. On the Surface Web, when you visit a website, your computer first deciphers what the IP address is for that website. As a result, the IP address can be linked to a business or individual. TOR hides the IP address of a website from its users. Such sites are called "hidden services" or "onion services" (Biryukov et al. 2014). In a rare study investigating TOR hidden services, Biryukov et al. (2014) exploited a temporary flaw in the software to estimate popularity among TOR services and found that the most frequently requested sites were pornographic (no information was provided on whether these sites depicted children or adults). A more recent study of TOR services found that 83% of TOR hidden service requests were for online destinations with child pornographic exploitive content. The requests of hidden services with child pornographic exploitive content completely dominated the number of requests for any of the other categories of services, such as purchase of illegal drugs, gambling, bitcoin-related sites, or anonymous whistle-blowing activity (Owen 2014). One limitation of this study is that the researchers were not able to determine how many of these requests for access to child pornographic sites were made by individuals, and how many might have been requested by Web-crawlers used by law enforcement to track traffic to these sites. Moore and Rid (2016) collected data regarding the types of hidden services on the Darknet through use of a website crawler. These researchers identified 122 hidden services related to illicit pornography and their descriptions of these services were disturbing:

[4] TOR. The TOR Project. https://blog.torproject.org/tor-project-elects-new-board-directors

[5] I2P, The Invisible Internet Project. https://geti2p.net/en/

[6] Freenet, The Free Network. https://freenetproject.org/index.html

> Websites dedicated to providing links to videos purporting to depict rape, bestiality and paedophilia were abundant. One such post at a supposedly nonaffiliated content-sharing website offered a link to a video of ‘a 12 year old girl … getting raped at school by 4 boys’…Other examples include a service that sold online video access to the vendor’s own family members: ‘My two stepsisters … will be pleased to show you their little secrets. Well, they are rather forced to show them, but at least that’s what they are used to’ (Moore and Rid 2016, p. 23).

Payment for illicit services on the Darknet can be made using cryptocurrency such as Bitcoin or prepaid debit cards, further enhancing protection for buyers due to the anonymity of both payment options. Bitcoin is a decentralized electronic payment system based on peer-to-peer networks instead of relying on central administration like banks. Bitcoin is designed as an open-source currency, where no single individual, corporation, or government owns or controls it. While Bitcoin is extremely difficult to trace and therefore considered an anonymous currency. However, as it is technically possible to link a user to an address by exploring transactions on a *blockchain,* Bitcoin is actually “pseudoanonymous” (for more information, see Conti et al. 2018). All financial transactions using Bitcoin are recorded on a blockchain, though the only information recorded are the user and sender addresses. These addresses do not provide the identity of the particular bitcoin, the user, or the sender – they only identify the specific transaction. Users and senders are linked to a “wallet”. Wallets contain private keys or secret numbers (similar in functionality to passwords) that allow individuals to spend bitcoins in the wallets. The wallets and private keys are not recorded in the public ledger, thereby, facilitating anonymity (Finklea 2017). This anonymity is of course advantageous to those committing illicit acts in the Darkweb, including human trafficking.

Evidence of the proliferation and lucrative earnings linked to child pornographic exploitation facilitated by the Darknet have surfaced through reports of arrests and prosecutions of those involved. One case involved the arrest of Peter Scully by Philippine investigators in the Anti-Human Trafficking Division of the Philippine National Bureau of Investigation for the production and dissemination of sadistic child sexual abuse videos sold through pay-per-view channels to an international pedophile ring operating on the Darknet. Scully allegedly produced a series of films titled *Daisy’s Destruction*, which involved the brutal torture, rape, and seeming murder of an 18-month-old toddler.

> Additional *Daisy’s Destruction* films include: a) a 5-year-old girl hung upside down while Scully and two accomplices raped and tortured her, b) two cousins, ages 9 and 12, forced to perform sex acts on each other, and, c) three children who were tortured, brutally raped, and made to appear murdered. The three children appeared to be murdered were: a) “Daisy,” age 18 months, who was later found with severe physical injuries, b) “Cindy,” age 11 years, who is alleged to have later been strangled to death by Scully and found buried under a house he rented in Malaybalay, Bukidnon, in the Philippines, and, c) “Liza,” age 12 years, who was found alive.” (Lacter 2018, para. 7–8).

Customers on the Darknet allegedly paid $10,000 each to view *Daisy’s Destruction* films (Lacter 2018). Scully was charged and convicted of human trafficking and rape, and he faces additional indictments for torture and murder.

4 Criminal Justice Response

During the last decade, there has been a growing awareness of and concerted efforts by law enforcement agencies at all levels – international, national, state, and local agencies – to combat human trafficking facilitated by digital technology and the Internet. Additionally, human trafficking cases frequently traverse jurisdictional lines making cooperation between local, state, federal, and even international law enforcement agencies a necessary component. Specialized cybercrime units and human trafficking task forces have countered this challenge by facilitating enhanced cooperation between federal, state, and local agencies. Multi-agency cooperation facilitated by these specialized units provides a convincing example of how various agencies within the criminal justice system can collaborate to address human trafficking.

The properties of the Surface Web and Darknet segments of the Internet provide law enforcement agencies with an electronic environment where they are able to conduct surveillance. For example, law enforcement agencies take advantage of the Internet's anonymity to conduct online "sting" operations by creating pseudo-personas to identify and arrest offenders. Law enforcement may pose as a potential victim or a fellow-offender in order to attract offenders and collect enough evidence for an arrest. Data mining tools assist law enforcement investigators to identify human traffickers. Using these technological advanced tools, law enforcement can scan extremely large data sets and use computers to find patterns, trends and associations related to human trafficking.

Law enforcement agencies engage in various online policing strategies to combat human trafficking facilitated through the Surface Web, Deep Web, or Darknet, including online sting operations, investigations of websites offering illegal content, and investigations arising from public complaints or cyber-tip hotlines (Krone 2005a). The following section describes how these strategies have been developed and used by law enforcement conducting human trafficking investigations.

Online sting operations resemble traditional sting operations. By definition, sting operations involve four elements:

1. an opportunity or enticement to commit a crime, either created or exploited by police.
2. a targeted likely offender or group of offenders for a particular crime type.
3. an undercover or hidden police officer or surrogate, or some form of deception.
4. a "gotcha" climax when the operation ends with arrests (Newman and Socia 2007, p. 3)

When conducting online sting operations, law enforcement agents create a presence online and pretend to be potential victims or peer-offenders in order to lure potential offenders into the commission of a crime (Fortin and Paquette 2018). Online stings are frequently used to combat two common forms of human trafficking – child pornography and commercial sexual exploitation of children. Most often, for stings related to these types of sexual offenses against children, officers pretend to be minors in online spaces and aim to identify persons attempting to meet offline with children (Krone 2005b; Mitchell et al. 2010).

Alternately, law enforcement agencies may use decoy or even commandeered web sites that contain or appear to contain child pornographic materials, in order to identify child pornography viewers and collectors (Riley 2017; Thompson 2003). In 2015, the FBI investigated and commandeered a child pornography website known as *Playpen*, which was operating on the Dark Web and had nearly 215,000 members. As part of the investigation, a Virginia District Court judge authorized a search warrant allowing the FBI to employ a network investigative technique (NIT) to try to identify actual IP addresses of computers used to access *Playpen*. Through the use of the NIT, the FBI was able to uncover about 1,300 IP addresses and subsequently trace those to individuals. Criminal charges were filed against more than 185 people. However, the results of the investigation faltered in 2016 when a judge ruled that all evidence collected during the time the FBI ran the site was invalid because of an incorrect warrant. Moreover, in a different ruling, the court found that malware used by the FBI to gather evidence about site users was unconstitutional because the FBI had failed to gain a search warrant to use it (Riley 2017).

Reviewing and monitoring of Internet ads is another common method of identifying human trafficking offenses, offenders, and victims (Heil and Nichols 2014). Frequently, these investigations are the result of tips from concerned caregivers or community members. In the United States, free websites such as backpage.com, the adult section of Craigslist, and Facebook have been identified as websites used by sex traffickers (Bouche and THORN 2015). By searching these and other websites, police have been able to identify hundreds of victims and traffickers (Heil and Nichols 2014). As part of the U.S. Federal Bureau of Investigation (FBI) Innocence Lost National Initiative, which began in 2003, Operation Cross Country has yielded more than 6,500 child identifications and locations (FBI 2017). As part of this operation, agents identified potential human trafficking victims through online ads for sexual services. Subsequently, undercover agents and detectives met with these potential human trafficking victims in order to identify the individuals who were exploiting them (FBI Seattle 2017).

As previously noted, the crime of human trafficking often involves more than one location or jurisdiction. Similarly, information related to human trafficking victims or traffickers is not limited to one website but rather is generally spread across various locations on the Internet. Researchers at DARPA found that over a 2-year period, sex traffickers in the United States spent in excess of $60 million to post 250 million advertisements on the Internet (Greenemeier 2015). Taking advantage of online advertising of sex trafficking on the Internet, data search and extraction systems (such as DeepDive[7]) have been developed and applied to identify human traffickers and human trafficking networks based on threads of information such as names, addresses, phone numbers, job types, job requirements, information about rates of service numbers, etc. posted repeatedly on the Internet. Using DeepDive[vii]

[7] DeepDive. http://deepdive.stanford.edu/

technology, DARPA developed the Memex[8] program, which has been used to trace human traffickers and assist prosecutors collect evidence to secure human trafficking convictions (Greenemeier 2015; Mensikova and Mattmann 2018). The developers of the Memex project used a portfolio approach to create a variety of "20 new super-search-engine parts, coded by 17 different units from private industry and universities, and dedicated to providing better ways of interacting with and understanding the data available on the whole Internet" (Graeber 2016, p. 11). Different tools[9] were developed to search and extract information from the Surface Web, the Deep Web, and Darknet. Tools, such as DIG[10] and TellFinder,[11] are available to law enforcement or specific research groups. TellFinder[xi] identifies and organizes repeated information from advertisements for sexual services. In the way, law enforcement investigators gain a clearer understanding of the scope of any identified enterprise. In a demonstration of TellFinder[xi], a news reporter was shown 869,000 current ads represented like population-density bubbles scattered across the United States (Graeber, 2016). Using the TellFinder[xi] tool, the analyst was able to zoom in on specific jurisdictions and scroll backward through posting dates, exposing where ads were posted over time – creating a chronological and geographical track of activity related to one enterprise or one specific person being trafficked (Graeber 2016).

Another common origin of human trafficking investigation are reports or tips from the public and advocacy groups. In one well-known example, a child advocacy agency in the Netherlands created a fake identity of a computer-generated child named "Sweetie" who was contacted by 20,000 men who requested Sweetie perform sex acts in front of a webcam for cash (Stone 2015; Terres des hommes 2013). The agency sent details of over 1,000 men to police agencies in 71 countries around the world; an Australian man, a registered sex offender, was the first to be convicted based on this information (Crawford 2013).

In conclusion, human trafficking has historically been a very difficult crime to identify and disrupt, due to the efforts to conceal its occurrence by traffickers, buyers/johns, and in many cases, the victims themselves. This challenge has only been exacerbated by the proliferation of the Darknet, anonymized currency, and efforts to "normalize" sexual exploitation, particularly of minors, in online communities run by buyers and distributors of sex, labor, organs, and child pornography.

These scenarios have posed substantial challenges to law enforcement investigators and prevention efforts, though several success cases have been noted. For instance, Operation Cross Country XI, a massive operation involving 55 FBI field offices, 78 state and local human trafficking task forces, and hundreds of local law enforcement agencies were able to leverage technology, undercover and

[8] Memex. https://memex.jpl.nasa.gov/

[9] A list of memex-related tools and their repository URLs. https://github.com/darpa-i2o/memex-program-index

[10] DIG, Domain-specific Insight Graphs. http://usc-isi-i2.github.io/dig/

[11] Tellfinder. http://usc-isi-i2.github.io/dig/

surveillance efforts, and inter-agency collaborations to successfully arrest 120 traffickers and recover 84 minor victims in the 2017 sting operation (Keiper and Chiaramonte 2019a). Efforts like these demonstrate the value, and necessity, for agencies to work together to effectively combat human trafficking in the modern era where the practice largely takes place online, where there are no jurisdictional borders and faces/evidence to easily associate with the illegal activity.

While traffickers will continue to innovate and find new and better ways to evade law enforcement's reach, particularly on the Dark Web, some have suggested that in addition to identifying and arresting traffickers, the best way to combat the trade is to "quell the demand" (Keiper and Chiaramonte 2019b). As stated by Pasco Sheriff's Office Corporal Alan Wilkett, "Historically, we've allowed the sex buyer to hide behind a mask of anonymity and actually call them a 'John. We don't even call them by their name, because we let them stay behind that mask. And the only way we're going to attack the supply side is by going after the demand. And that means the sex buyer needs to be held culpable for the damage and trauma that he or she is causing" (Keiper and Chiaramonte 2019b).

In line with this sentiment, prevention efforts focusing on reducing demand by illustrating to buyers the severe damage that trafficking causes to victims and increasing accountability, particularly for repeat buyers, appear to be promising avenues to reduce human trafficking on- and off-line. However, as the United States ranks as one of the worst countries in the world for human trafficking, in terms of origin nation for victims (U.S. State Department 2019), far more must also be done to identify, recover, and treat trafficking survivors. As the U.S. Department of Justice recently increased funding to $31 million in 2018 for government and non-profit organizations aimed at providing services to trafficking survivors, up from $16 million in 2017, there is hope that those most impacted by this despicable offense will receive more aid, and be made whole for the crimes they experienced. As researchers, practitioners, and policymakers, it is up to us to do our part to help combat this crime and its expanding domain in the digital world.

References

Açar, K. V. (2017). Organizational aspect of the global fight against online child sexual abuse. *Global Policy, 8*(2), 259–262.

Aked, S. (2011). An investigation into darknets and the content available via anonymous peer-to-peer file sharing. In *Proceedings of the 9th Australian information security management conference, Edith Cowan University, Perth Western Australia, 5th – 7th December*. Retrieved from http://ro.ecu.edu.au/ism/106

Armagh, D. S., Battaglia, N. L., & U.S Department of Justice. (2006). *Use of computers in the sexual exploitation of children*, 2nd ed. Washington, DC.

Bartlett, J. (2014). *The dark net: Inside the digital underworld*. London: William Heinemann.

Beech, A. R., Elliott, I. A., Birgden, A., & Findlater, D. (2008). The internet and child sexual offending: A criminological review. *Aggression and Violent Behavior, 13*(3), 216–228. https://doi.org/10.1016/j.avb.2008.03.007.

Bergman, M. K. (2001). White paper: The deep web: Surfacing hidden value. *Journal of Electronic Publishing, 7*(1). Retrieved from https://quod.lib.umich.edu/cgi/t/text/idx/j/jep/3336451.0007.104/%2D%2Dwhite-paper-the-deep-websurfacing-hidden-value?rgn=main;view=fulltext.

Biryukov, A., Pustogarov, I., Thill, F., & Weinmann, R. P. (2014, June). Content and popularity analysis of Tor hidden services. In *2014 IEEE 34th International Conference on Distributed Computing Systems Workshops (ICDCSW)* (pp. 188–193). IEEE.

Bouche, V. & THORN. (2015). *A report on the use of technology to recruit, groom and sell domestic minor sex trafficking victims*. Technical Report, THORN. Retrieved from https://www.wearethorn.org/wp-content/uploads/2015/02/Survivor_Survey_r5.pdf

Brennan, D. (2018, November 7). Former police chief arrested in baby trafficking ring bust. *Newsweek*. Retrieved from https://www.newsweek.com/former-police-chief-arrested-baby-trafficking-ring-bust-1205404

Çalışkan, E., Minárik, T., & Osula, A.-M. (2015). *Technical and legal overview of the tor anonymity network*. NATO Cooperative Cyber Defense Centre of Excellence. Retrieved from https://www.ccdcoe.org/library/publications/technical-and-legal-overview-of-the-tor-anonymity-network/

Caulfield, T., Duijst, W., Bos, M., Chassis, I., Codreanu, I., Danovitch, G., et al. (2016). Trafficking in human beings for the purpose of organ removal and the ethical and legal obligations of healthcare providers. *Transplantation direct, 2*(2), e60–e60. https://doi.org/10.1097/TXD.0000000000000566.

Conti, M., Kumar, E. S., Lal, C., & Ruj, S. (2018). A survey on security and privacy issues of bitcoin. *IEEE Communications Surveys & Tutorials, 20*(4), 3416–3452. https://doi.org/10.1109/COMST.2018.2842460.

Crawford, A. (2013, November 5). Computer-generated 'Sweetie' catches online predators. *BBC News*. Retrieved from https://www.bbc.com/news/uk-24818769

Crawford, A. (2016, August 12). Paedophiles use secret Facebook groups to swap images. *BBC News*. Retrieved from https://www.bbc.com/news/uk-35521068

Delmonico, F. L. (2009). The implications of Istanbul Declaration on organ trafficking and transplant tourism. *Current Opinion in Organ Transplantation, 14*(2), 116–119. https://doi.org/10.1097/MOT.0b013e32832917c9.

Federal Bureau of Investigation (FBI). (2017). *FBI announces results of operation cross country XI: Underage sex trafficking crackdown leads to recovery of 84 minors*. Retrieved from https://www.fbi.gov/news/pressrel/press-releases/fbi-announces-results-of-operation-cross-country-xi

Federal Bureau of Investigation (FBI) Seattle. (2017). *Eleventh FBI-wide operation cross country arrests three suspected sex traffickers in Washington state*. Retrieved from https://www.fbi.gov/contact-us/field-offices/seattle/news/press-releases/11th-fbi-wide-operation-cross-country-arrests-three-suspected-sex-traffickers-in-washington-state

Finklea, K. (2017, March 10). *Dark Web*. Congressional Research Service CRS Report. Retrieved from https://fas.org/sgp/crs/misc/R44101.pdf

Finn, M. A., & Stalans, L. J. (2016). How targeted enforcement shapes marketing decisions of pimps: Evidence of displacement and innovation. *Victims & Offenders, 11*(4), 578–599. https://doi.org/10.1080/15564886.2016.1187694.

Florida no haven for teen exploitation. (2008, June 6). Editorial. *The Tampa Tribune (Tampa, FL)*, p. 2. Retrieved from http://link.galegroup.com/apps/doc/A181350103/ITOF?u=tamp44898&sid=ITOF&xid=bef19b52

Fortin, F., & Paquette, S. (2018). Online Sexual Exploitation of Children: Reactive and proactive policing. In P. Lussier & E. Beauregard (Eds.), *Sexual Offending: A Criminological Perspective* (pp. 237–256). New York, NY: Routledge.

Global Financial Integrity. (2017). *Transnational crime and the developing world*. Retrieved from http://www.gfintegrity.org/wp-content/uploads/2017/03/Transnational_Crime-final.pdf

Gonzales, A. R. (2006). *Transcript of attorney general Alberto R. Gonzales' address to the employees at the national center for missing and exploited children*. Washington, DC, April 20. Retrieved from http://www.justice.gov/archive/ag/speeches/2006/ag_speech_0604202.html

Graeber, C. (2016). The man who lit the Dark Web: Data-mining tools are helping cops bust open online human trafficking. *Popular Science*. Retrieved from https://www.popsci.com/man-who-lit-dark-web/

Greenemeier, L. (2015, February 8). Human traffickers caught on hidden internet. *Scientific American*. Retrieved from https://www.scientificamerican.com/article/human-traffickers-caught-on-hidden-internet/

Gupta, J. A. (2018). Reproductive labour or reproductive trafficking?: Indian women's reproductive bodies in the globalised bioeconomy. In R. Ryan-Flood & J. G. Payne (Eds.), *Transnationalising reproduction* (pp. 95–116). Routledge.

Heil, E., & Nichols, A. (2014). Hot spot trafficking: A theoretical discussion of the potential problems associated with targeted policing and the eradication of sex trafficking in the United States. *Contemporary Justice Review, 17*(4), 421–433.

Janson, L., Durchslag, R., Mann, H., Marro, R., & Matvey, A. (2013). *"Our great hobby": An analysis of online networks for buyers of sex in Illinois*. Chicago Alliance Against Sexual Exploitation (CAASE). Retrieved from https://humantraffickinghotline.org/sites/default/files/Our%20Great%20Hobby%20-%20CAASE%20and%20Thorn_0.pdf

Johnson, R. (2019). *How the U.S. immigration system encourages child marriages*. United States. Congress. Senate. Committee on Homeland Security and Governmental Affairs. Retrieved from https://www.hsdl.org/?view&did=820021

Jones, J. (2011). Trafficking internet brides. *Information & Communications Technology Law, 20*(1), 19–33.

Keiper, A., & Chiaramonte, P. (2019a, June 1). The 'village' of law enforcement agencies battling human trafficking. *Fox News Investigates*. Retrieved from https://www.foxnews.com/us/the-village-of-law-enforcement-agencies-battling-human-trafficking

Keiper, A., & Chiaramonte, P. (2019b, June 22). Human trafficking in America among worst in world: Report. *Fox News Investigates*. Retrieved from https://www.foxnews.com/us/human-trafficking-in-america-among-worst-in-world-report

Klain, E. J., Davies, H. J., & Hicks, M. A. (2001). *Child pornography: The criminal-justice-system response*. National Center for Missing & Exploited Children.

Krone, T. (2005a). International police operations against online child pornography. *Crime & Justice International, 21*(89), 1–6.

Krone, T. (2005b). *Queensland police stings in online chat rooms*. Australian Institute of Criminology. Retrieved from https://apo.org.au/sites/default/files/resource-files/2005/10/apo-nid2358-1095206.pdf

Lacter, E. P. (2018, September 29). *Child rape and torture materials on the dark web: Peter Scully and beyond*. Retrieved from https://endritualabuse.org/child-rape-torture-materials-dark-web/

Latonero, M. (2011). *Human trafficking online: The role of social networking sites and online classifieds*. Available at SSRN 2045851.

Latonero, M., Musto, J., Boyd, Z., Boyle, E., Bissel, A., Gibson, K., & Kim, J. (2012). *The rise of mobile and the diffusion of technology-facilitated trafficking* (p. 43). Los Angeles: University of Southern California, Center on Communication Leadership & Policy.

Latonero, M., Wex, B., Dank, M., & Poucki, S. (2015). *Technology and labor trafficking in a network society*. USC Annenberg Center on Communication Leadership and Policy, Report. Retrieved from https://communicationleadership.usc.edu/files/2015/10/USC_Tech-and-Labor-Trafficking_Feb2015.pdf

Mensikova, A., & Mattmann, C. A. (2018). *Ensemble sentiment analysis to identify human trafficking in web data*. NASA. Retrieved from https://memex.jpl.nasa.gov/GTA2018-Sentiment.pdf

Mitchell, K. J., Finkelhor, D., Jones, L. M., & Wolak, J. (2010). Growth and change in undercover online child exploitation investigations, 2000–2006. *Policing and Society, 20*(4), 416–431.

Moore, D., & Rid, T. (2016). Cryptopolitik and the Darknet. *Survival, 58*(1), 7–38.

Murphy, L., Ranger, R., & Fedoroff, J. P. (2014). Legal and clinical issues in interpreting child pornography on the internet. In F. M. Saleh, A. Grudzinskas, & A. Judge (Eds.), *Adolescent sexual behavior in the digital age: Considerations for clinicians, legal professionals, and educators* (pp. 215–243). New York: Oxford University Press.

Newman, G. R., & Socia, K. (2007). Sting operations. Washington, DC: U.S. Department of Justice, Office of Community Oriented Policing Services.

Nwaka, J. C., & Odoemene, A. (2019). "Baby factories": Exploitation of women in Southern Nigeria. *Dignity: A Journal on Sexual Exploitation and Violence, 4*(2), 2. https://doi.org/10.23860/dignity.2019.04.02.02.

Owen, G. (2014, December 30). *Tor: Hidden services and deanonymisation*. Presentation #6112 at the Chaos Communication Congress, Hamburg, Germany. Retrieved from https://media.ccc.de/v/31c3_-_6112_-_en_-_saal_2_-_201412301715_-_tor_hidden_services_and_deanonymisation_-_dr_gareth_owen

Pugliese, E. (2007). Organ trafficking and the TVPA: Why one word makes a difference in international enforcement efforts. *Journal of Contemporary Health Law & Policy, 24*, 181–208.

Quayle, E., & Jones, T. (2011). Sexualized images of children on the internet. *Sexual Abuse: A Journal of Research and Treatment, 23*(1), 7–21.

Quayle, E., & Taylor, M. (2002). Child pornography and the internet: Perpetuating a cycle of abuse. *Deviant Behavior, 23*(4), 331–361.

Reid, J. (2008). *Rapid assessment of domestic minor sex trafficking in the Clearwater/Tampa Bay area*. Retrieved from http://www.northeastern.edu/humantrafficking/wp-content/uploads/Clearwater.pdf

Reid, J. A. (2015). Sex trafficking of girls: How sex traffickers exploit youth and system vulnerabilities. In *Paper presented at the anti-trafficking initiative lecture series*. St. Louis: Washington University in St. Louis. Retrieved from https://digital.usfsp.edu/fac_publications/217/

Reid, J. A. (2016a). Entrapment and enmeshment schemes used by sex traffickers. *Sexual Abuse, 28*(6), 491–511. https://doi.org/10.1177/1079063214544334.

Reid, J. A. (2016b). Child pornography in the 21st century: From child pornographic exploitation to youth sexting. *Oxford Handbooks Online*. Oxford University Press. https://doi.org/10.1093/oxfordhb/9780199935383.013.132.

Reid, J. A. (2018). System failure! Is the Department of Children and Families (DCF) facilitating sex trafficking of foster girls? In A. G. Nichols, T. Edmond, & E. C. Heil (Eds.), *Social work practice with survivors of sex trafficking and commercial sexual exploitation* (pp. 296–315). New York: Columbia University Press.

Riley, D. (2017, October 8). *Australian police ran a child porn site for nearly a year to catch pedophiles*. SiliconANGLE Media Inc. Retrieved from https://siliconangle.com/2017/10/08/australian-police-hosted-ran-child-porn-site-nearly-year-catch-pedophiles/

Rodas, E. (2014). *The multi-facets of cyber-sex trafficking*. A Call for Action and Reform from Society. GRIN Verlag.

Scheper-Hughes, N. (2003). Rotten trade: Millennial capitalism, human values and global justice in organs trafficking. *Journal of Human Rights, 2*(2), 197–226.

Stone, J. (2015, March 19). 'Sweetie 2.0' software uses virtual girl to lure child porn offenders out of the dark. *International Business Times*. Retrieved from https://www.ibtimes.com/sweetie-20-software-uses-virtual-girl-lure-child-porn-offenders-out-dark-1848724

Terre des hommes. (2013). *Webcam child sex tourism: Becoming sweetie: A novel approach to stopping the global rise of webcam child sex tourism.* Retrieved from http://www.terredeshommes.org/wpcontent/uploads/2013/11/FAQ-English.pdf

Thompson, T. (2003). Net closes on British paedophiles. *The Observer UK*. Retrieved from https://www.theguardian.com/society/2003/aug/17/epublic.childrensservices

Trafficking Victims Protection Act (TVPA) of 2000, 22 U.S.C. x7105(b)(1)(E)(u)[2000].

Turner, L. (2008). 'Medical tourism' initiatives should exclude commercial organ transplantation. *Journal of the Royal Society of Medicine, 101*(8), 391–394.

U.S. Department of Homeland Security. (2019). *El Salvador national charged with posting child pornography on Facebook and illegally reentering the US after deportation*. Retrieved from https://www.ice.gov/news/releases/el-salvador-national-charged-posting-child-pornography-facebook-and-illegally

U.S. Department of Justice [DOJ]. (2013, March 1). *California man arrested in federal sex trafficking case* [Press release]. Retrieved from https://www.justice.gov/opa/pr/california-man-arrested-federal-sex-trafficking-case

U.S. Department of State. (2019, June 20). *2019 trafficking in persons report*. Washington, DC: U.S. Department of State. Retrieved from https://www.state.gov/wp-content/uploads/2019/06/2019-Trafficking-in-Persons-Report.pdf

U.S. Immigration and Customs Enforcement. (2015). *Georgia man sentenced to more than 21 years for sexually exploiting minors in Thailand*. Retrieved from http://www.ice.gov/news/releases/georgia-man-sentenced-more-21-years-sexually-exploiting-minors-thailand

United Nations Office on Drugs and Crime [UNODC]. (2015a). *The role of recruitment fees and abusive and fraudulent recruitment practices of recruitment agencies in trafficking in persons*. Retrieved from https://www.unodc.org/documents/human-trafficking/2015/Recruitment_Fees_Report-Final-22_June_2015_AG_Final.pdf

United Nations Office on Drugs and Crime [UNODC]. (2015b). *Assessment_Toolkit_TIP_for_the_Purpose_of_Organ_Removal*. Retrieved from https://www.unodc.org/documents/human-trafficking/2015/UNODC_Assessment_Toolkit_TIP_for_the_Purpose_of_Organ_Removal.pdf

Wolak, J., Finkelhor, D., & Mitchell, K. (2011). Child pornography possessors: Trends in offender and case characteristics. *Sexual Abuse: A Journal of Research and Treatment, 23*(1), 22–42.

Wolak, J., Liberatore, M., & Levine, B. N. (2014). Measuring a year of child pornography trafficking by U.S. computers on a peer-to-peer network. *Child Abuse & Neglect, 38*(2), 347–356.

Wortley, R. K., & Smallbone, S. (2006). *Child pornography on the internet*. US Department of Justice, Office of Community Oriented Policing Services.

Wortley, R., & Smallbone, S. (2012). *Internet child pornography: Causes, investigation, and prevention*. ABC-CLIO.

Part II
Policing Tools & Strategies

Scientific and Technological Advances in Law Enforcement Intelligence Analysis

Randy Borum

Abstract Powered by advances in computing technology, a range of professions and business enterprises have moved toward a more science-driven approach to operations. Law enforcement has been no exception. In fact, the modern day idea of intelligence-led policing (ILP) emerged in the UK in the 1990s as the country was pushing all government services to operate on more of a data-informed, business process or managerial model. This trend led to the development of a British "National Intelligence Model" (NIM), which by 2002, was formally adopted as policy for law enforcement agencies nationwide. "The NIM followed the government policy of using a business process model to deal with crime control and employed the ILP philosophy to introduce intelligence into virtually all aspects of the policing business plan" (p. 311).

ILP has surged in popularity among US law enforcement agencies, although what exactly ILP means in an operational sense, and how it is implemented varies considerably (Ratcliffe J, Intelligence-led policing. Routledge, New York, 2016). At the most basic level, ILP is commonly defined as "the collection and analysis of information related to crime and conditions that contribute to crime, resulting in actionable intelligence products intended to aid law enforcement in developing tactical responses to threats and/or strategic planning related to emerging or changing threats" (Carter DL, Carter JG, Crim Justice Policy Rev 20(3):310, 2009, p. 317). Although sometimes broadly considered, ILP has become widely used. It is not a "program," but more of a philosophy and approach for meeting a law enforcement agency's mission objectives. This chapter will not focus primarily on the programmatic aspects of ILP, but on scientific and technological advances that have enhanced and accelerated the intelligence analysis process for policing applications.

Keywords Intelligence led policing · National intelligence model · Law enforcement · Data

R. Borum (✉)
School of Information, University of South Florida, Tampa, FL, USA
e-mail: borum@usf.edu

B. Fox et al. (eds.), *Science Informed Policing*, Advanced Sciences and Technologies for Security Applications,
https://doi.org/10.1007/978-3-030-41287-6_6

1 Introduction

Powered by advances in computing technology, a range of professions and business enterprises have moved to toward a more science-driven approach to operations. Law enforcement has been no exception. In fact, the modern day idea of intelligence-led policing (ILP) emerged in the UK in the 1990s as the country was pushing all government services to operate on more of a data-informed, business process or managerial model. This trend led to the development of a British "National Intelligence Model" (NIM), which by 2002, was formally adopted as policy for law enforcement agencies nationwide. "The NIM followed the government policy of using a business process model to deal with crime control and employed the ILP philosophy to introduce intelligence into virtually all aspects of the policing business plan" (p. 311).

ILP has surged in popularity among US law enforcement agencies, although what exactly ILP means in an operational sense, and how it is implemented varies considerably (Ratcliffe 2016). At the most basic level, ILP is commonly defined as "the collection and analysis of information related to crime and conditions that contribute to crime, resulting in an actionable intelligence products intended to aid law enforcement in developing tactical responses to threats and/or strategic planning related to emerging or changing threats" (Carter and Carter 2009, p. 317). Although sometimes broadly considered, ILP has become widely used. It is not a "program," but more of a philosophy and approach for meeting a law enforcement agency's mission objectives.

This chapter will not focus primarily on the programmatic aspects of ILP, but on scientific and technological advances that have enhanced and accelerated the intelligence analysis process for policing applications. We begin by defining intelligence and differentiating it from crime "data" or crime-related information, then we briefly discuss progress in the professionalization of law enforcement intelligence as a field of practice. Next, we demonstrate how science has helped advance the "human" dimensions of analytic tradecraft. Finally, we review recent developments in computing and in information and communication technology that have enabled analytic advances for policing applications.

1.1 Intelligence, Data and Information

Too often, the term "intelligence" is used casually, even in law enforcement parlance, to refer to crime data, statistics, or any potentially useful points of crime-related information. Intelligence may build on data or information, but the terms and concepts are not synonymous. In the field of Information Science there is a widely acknowledged hierarchy of information-related content, centered on four concepts: data, information, knowledge and wisdom (DIKW) (Ackoff 1989). Sometimes "understanding" is considered a separate category between knowledge and wisdom. Each cumulatively builds upon the other. In fact, this particular stack is often called the DIKW Hierarchy or just the "Information Hierarchy" (Rowley 2007). Systems theorist, Russell Ackoff (1989), defined the elements of this construct in the following way:

- Data: symbols with no inherent meaning
- Information: data that have been given meaning; provides answers to "who", "what", "where", and "when" questions
- Knowledge: application of data and information; answers "how" questions
- Understanding: appreciation of "why"
- Wisdom: evaluated understanding.

In this hierarchy, information is defined in terms of data, knowledge is defined in terms of information, and wisdom is defined in terms of knowledge (Rowley 2007). Analysis is the mechanism to get from data to information and from information to knowledge, understanding, and wisdom.

Intelligence, akin to "wisdom" in the Information Hierarchy, is a high-level concept. It is not just "processed" information; rather intelligence refers to the actionable insights derived from analyzed and evaluated information. As intelligence expert Robert Clark (2019) suggests: "A typical goal of intelligence is to establish facts and then to develop precise, reliable, and valid inferences (hypotheses, estimations, conclusions, or predictions) for use in strategic decision-making or operational planning" (p. 22). In that sense, intelligence is information in action, or in Michael Warner's (2002) words, "It is what people do with data and information that gives them the special quality we casually call intelligence" (p. 17).

Concerning Law Enforcement Intelligence, specifically, Carter (2009) offers the following insightful definition:

> intelligence is the end product of an analytic process that evaluates information collected from diverse sources; integrates the relevant information into a logical package; and produces a conclusion, estimate, or forecast about a criminal phenomenon by using the scientific approach to problem solving (that is, analysis). Intelligence, therefore, is a synergistic product intended to provide meaningful and trustworthy actionable knowledge to law enforcement decision makers about complex criminality, criminal enterprises, criminal extremists, and terrorists (p. 9).

Data, information and intelligence have all supported law enforcement operations, but the strategic and tactical uses of intelligence have created a modern ethos of "evidence based policing." The status of intelligence in American law enforcement was markedly elevated in the wake of the 9/11 terrorist attacks. Those attacks were widely cited as resulting from a "failure of intelligence." In the post-9/11 era, federal, state, local and tribal law enforcement agencies were pushed to move from traditional approaches of response driven policing to more proactive approaches emphasizing prevention and risk management. They were also given a much more central role (and responsibility) for homeland defense and security. Improvements and advancements in law enforcement intelligence were often the centerpiece of those efforts.

Because intelligence is defined by the meaning educed from processed and analyzed information, all intelligence, including law enforcement intelligence, emerges from a recursive sets of tasks and activities, known as the Intelligence Cycle. The traditional intelligence cycle comprises the following five stages:

- **Planning and Direction**: The first stage involves establishing what intelligence the consumer/requestor needs to best inform her/his decision. The planner should identify not only she/he thinks the customer "needs to know," but what the consumer will *do* with the information. The intelligence points needed for the decision are known as "Intelligence Requirements." The requirements guide the plan for what data or information will be collected (as well as how and from where it will be collected), what kind of analysis or processing is needed, and to whom (and how) the finished intelligence product it will be disseminated.
- **Collection**: In the second stage, the raw data or information is collected in accordance with the plan. That data may come from publicly available information (Open Source) and/or from secret or classified sources, and may be gathered from straightforward searches of textual materials, queries of human assets, or through technical means (e.g., electronic, satellite imagery, measurements, etc).
- **Processing**: In the third stage, the raw data or information is processed into a usable form. That may involve decryption, language translations, technical measurement or statistical analysis. In the era of Big Data and information overload, a critical preliminary task for the processing stage may be data reduction – sifting through the "noisy" data and drilling down to the most vital and reliable points. This might be regarded as a type of "pre-processing."
- **Analysis and Production**: The fourth stage is where data becomes intelligence. An analyst evaluates and integrates all relevant data/information from various sources, attempts to "make sense" of it, and to draw inferences and conclusions with regard to the original intelligence requirements.
- **Dissemination**: In the fifth and final stage, the analytic assessment, a finished product, is provided to the intelligence consumer/requestor in a form that directly addresses the specified intelligence requirements. That will lead the intelligence consumer/requestor to determine whether the intelligence provided is sufficient, or whether additional intelligence requirements need to be prosecuted, which is the function of the first stage; hence, the cyclical nature of the process.

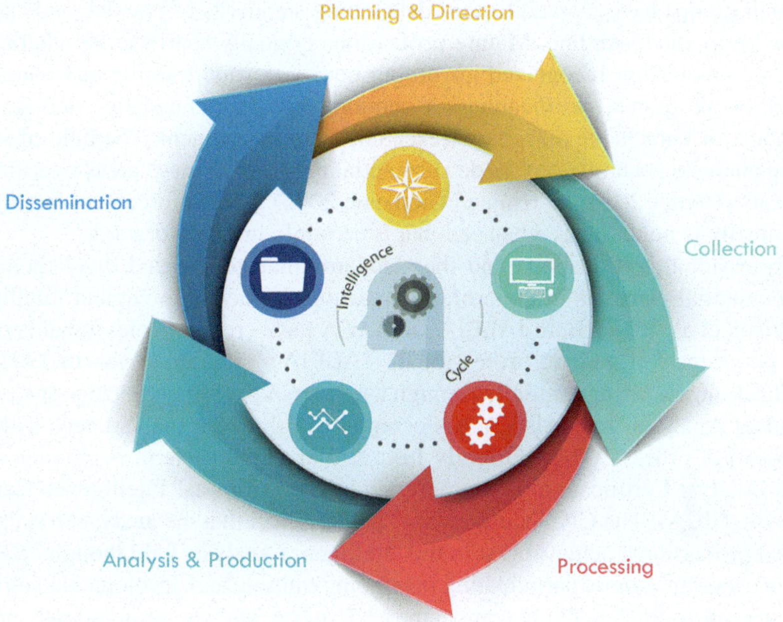

1.2 *Human-Centered Advances in Intelligence*

After 2001, as law enforcement agencies began building or growing their intelligence capabilities, demand grew quickly for qualified law enforcement intelligence analysts. Two related trends are especially noteworthy: (1) professionalization of law enforcement intelligence as a discipline and (2) the application of scientific knowledge to improve human analytic performance.

1.2.1 Professionalization of Law Enforcement Intelligence Analysis

Although intelligence and crime analysts have worked in law enforcement for decades, as a discipline, law enforcement intelligence analysis is still fairly young (Gentry 2016; Marrin 2016; Marrin and Clemente 2006). It was not until the 1980s that crime and intelligence analysts began networking, organizing and developing a collective sense of professional identity. The first professional association for law enforcement analysts— the International Association of Law Enforcement Intelligence Analysts, Inc. (IALEIA)—only began in October, 1981. They currently have over 2500 members from more than 40 countries. A second organization emerged in 1990—the International Association of Crime Analysts (IACA). They currently claim over 4000 active members from more than 60 countries.

Both groups have worked toward developing standards of practice and "professionalizing" the discipline. Mature professions typically share certain characteristics, such as having recognized governing bodies, formal training and continuing education programs, certifications, a common body of knowledge and standard methods for knowledge discovery (research) and management. The field of intelligence analysis generally has made substantial progress in these areas over the past 40 years (Gentry 2016; Marrin and Clemente 2006). Law enforcement intelligence specifically is not quite as advanced, but is certainly moving forward.

Professional certifications and the standards that govern and drive them have been essential elements in the professionalization of law enforcement intelligence (Harrison et al. 2018). Both IALEIA and IACA have created professional certification programs for analytic professionals. IALEIA's program began in 1991. The certification was crafted initially through a partnership with the Society of Certified Criminal Analysts, but IALEIA has essentially subsumed that Society within its certification process. They currently have a four tiered system, beginning with a Basic Analyst Certification, and progressing to the Criminal Intelligence Certified Analyst (CICA). The CICA designation should reflect that the analyst "has demonstrated professional achievement in the intelligence analysis field though establishing an experience base, participating in ongoing training and development activities, and demonstrating analytic competency." Criteria include professional analytic experience, specialized analytic training and formal education concerning 17 content/skill areas.

After 9/11, The International Association of Chiefs of Police (IACP) and the Community Oriented Policing Services (COPS) convened the "Summit on Criminal Information Sharing: Overcoming Barriers to Enhance Domestic Security." A central conclusion from that Summit was that standards were needed for intelligence training. This call for professional standards was echoed in *The IACP Criminal Intelligence Sharing Report: A National Plan for Intelligence-Led Policing at the Local, State and Federal Level*, and the Summit recommendations were formally adopted by The Global Justice Information Sharing Initiative (Global) Intelligence Working Group (GIWG) Training Committee. The GIWG then set about reviewing and establishing best practices in law enforcement intelligence from federal, state, and local-level programs.

In October 2003, the Global Justice Information Sharing Initiative released a *National Criminal Intelligence Sharing Plan* (NCISP), and in 2005 the GIGWG, in cooperation with IALEIA, released a set of 25 *Law Enforcement Analytic Standards* for intelligence analyst certification programs and processes. The standards pertained to analyst education/training, professional conduct, and certification as well as standards for the analytic processes and products themselves. Concerning professional certification, the *Standards* made the following specific recommendation:

> *Analysts should be certified by an agency or organization (governmental, professional association, or institution of higher learning) program specifically developed for intelligence analysts. These analytic certification programs shall reflect experience, education, training, and proficiency testing* (Global Justice Information Sharing Initiative 2007, p. 1).

This call accelerated the field's movement toward widespread endorsement of analyst certification (Global Justice Information Sharing Initiative 2003).

IALEIA's certification was aligned with NCISP guidelines. Building on the NCISP mandate, IACA's professional certification program—the Certified Law Enforcement Analyst (CLEA)—began testing analysts in 2005. Criteria for the CLEA credential also include professional analytic experience, specialized analytic training and formal education to include 19 content/skill areas. Both the CICA and CLEA certifications have authoritative reference documents comprising a common body of knowledge for law enforcement analytic practitioners.

A comparison of the Certification domains for the CICA and CLEA certifications in provided in Table 1.

The experience, training/education and professional knowledge requirements for the CICA and CLEA credentials are substantially similar. IALEIA tends to focus more on strategic-level analysis (e.g., longer term, future-oriented assessments), while IACA tends to emphasize the tactical-level (e.g., immediate and short-term characterizations of the operating environment). Similarly, IALEIA tends to highlight more traditional, qualitative, expert-driven assessment methods (with roots in structured analytic techniques), while IACA has a more explicit focus on quantitative, data-driven methods (with roots in crime statistics and analytics). Law enforcement analysts in both organizations (and with both certifications), however, can

Table 1 Comparison of IALEIA and IACA knowledge domains

IALEIA's CICA exam domains	IACA's CLEA exam domains
Analytic techniques and methods	Crime analysis – terms, concepts and processes
Creative and critical thinking	Law enforcement models
Crime indicators	Criminal behavior
Crime pattern analysis	LE data and crime analysis data sources
Inference development	Internet resources
Information management – classifications, collation	Crime analysis data and data management
Information sharing	Applied research methods
Information sources	Descriptive statistics
Intelligence cycle/process	Advanced statistical concepts
Intelligence-led policing	Qualitative analysis
Intelligence product development	Spreadsheet operations
Legal, privacy, ethics issues	Temporal analysis
Logic	Intelligence analysis and intelligence charting
Recommendation development	GIS and crime mapping
Structured analytic techniques	Spatial analysis and forecasting
Strategic analysis	Effective analytical writing
Tactical analysis	Analytical products
	Applied crime series analysis
	Organized crime

operate across the strategic-tactical spectrum and may use qualitative and/or quantitative analytic methods; the difference is mainly a matter of emphasis. Both organizations and both professional certifications have helped to standardize and raise the bar for analyst qualifications, just as the *Law Enforcement Analytic Standards* and *Law Enforcement Analyst Certification Standards* have elevated law enforcement analytic practices, processes and products (Bruce and George 2015).

1.2.2 Applying Science to Improve Human Analytic Performance

Intelligence analysis relies on complex cognitive processes, to include evaluating and weighing evidence to test hypotheses and educe conclusions under conditions of uncertainty. There are well-known limitations to those cognitive capabilities, however, that can directly affect analytic outcomes. Two applications of science (i.e. the science of human judgment and decision making) have had a particular impact on human analytic performance in intelligence-related analytic tasks. The first occurred over several years as a senior CIA analyst sought to educate analysts on decision-making research, and look for ways to mitigate the impact of cognitive biases. The second was a large-scale program of research aimed at understanding how people make forecasts and predictions about geopolitical events, and how those assessments might be optimized.

The first of these watershed applications began in the late 1970s and continued through the 1980s, as the Central Intelligence Agency's (CIA) Directorate of Intelligence (the Agency's analytic arm) began circulating a series of internal documents "reviewing cognitive psychology literature concerning how people process information to make judgments on incomplete and ambiguous information" (Heuer 1999, p. vii). The research findings were digested and illustrated with challenges facing intelligence analysts, and accompanying recommendations for how analysts might apply evidence-based strategies to mitigate these common cognitive limitations. Many of these documents were authored by Richards Heuer, who subsequently compiled and supplemented them to create a now-classic work titled, *Psychology of Intelligence Analysis*. That body of work may represent the origins of "analytic science."

Heuer's memos described research illuminating a number of cognitive biases, limitations and heuristics (mental "shortcuts") that are common in nearly all forms of human judgment and decision-making. Among the points revealed in studies were:

- The human brain is "wired" to look for cause-and-effect explanations. As a result, people develop their own "mental models" of the problems they are trying to solve. Those models carry causal assumptions that may or may not have any basis in fact. Sometimes people are aware of their implicit models and assumptions; sometimes not. But those models and assumptions guide their judgments.
- Although people often think that having more information always leads to better decisions, studies suggest that is not true. There are limits to the value (or

"incremental validity") of adding new information. In fact, by always seeking "more information" people can add valueless inputs that ultimately undermine the accuracy of their decisions.

- When people are looking for answers or solutions under conditions of uncertainty, they tend to look for, and therefore find, information that confirms their prior assumptions and previously held beliefs. Information that does not fit with what they already believe is typically overlooked or dismissed. This creates an inherent "confirmation bias" in their judgments.
- Although people (including analysts) tend to value the forecasts of "experts" in a field much more than the predictions of people who do not have subject matter expertise, experts rarely produce more accurate predictive judgments—and sometimes they do significantly worse than non-experts. They certainly have no greater immunity to the general cognitive limitations and biases that routinely plague human decision making.

There are dozens of other known cognitive biases and heuristics that can affect decision making. The bad news is that it is probably impossible to eliminate them. The good news is, with awareness of these potential cognitive hazards, it may be possible to mitigate them or reduce their impact by changing the way a problem is structured or solved. The key implication, as Heuer points out, is that "Intelligence analysts should be self-conscious about their reasoning processes. They should think about *how* they make judgments and reach conclusions, not just about the judgments and conclusions themselves" (Heuer 1999, p. 31, emphasis in original). The highlights from these studies are still included in most intelligence analysis training programs.

Heuer's persistent interest in reducing bias and improving analytic judgments, led him (and other analysts, such as Randy Pherson) to develop a set of "Structured Analytic Techniques" (SATs). These are tools and techniques to help an analyst expose and explore her/his implicit mental model, to explore alternatives that may not comport with his/her prior beliefs, and to challenge her/his own conclusions and underlying arguments (Heuer and Pherson 2010). Although a given technique may serve multiple functions, they are generally grouped into three categories: *Diagnostic Techniques* aim to illuminate assumptions and make judgments more transparent; *Contrarian Techniques* challenge current thinking; and *Imaginative Thinking Techniques* are used to encourage new insights, perspectives or possible outcomes.

These techniques have become an essential part of intelligence analytic tradecraft taught within the U.S. Intelligence Community and in most analytic training programs. The SATs also cohere with and support the U.S. Intelligence Community's Analytic Standards, as outlined in Intelligence Community Directive 203 (ICD 203). The Analytic Tradecraft Standards described in ICD 203 require, among other things, that analytic products describe the underlying sources and methods used for the analysis; explicitly address any uncertainties; distinguish source information from assumptions and analytic judgments; and incorporate an analysis of alternatives. SATs hold substantial promise for improving the quality of analytic judgments and products. Unfortunately, despite SATs widespread inclusion in training and policy, very little empirical research has been conducted to which ones work or when and how they may work (Artner et al. 2016).

The second major application of science to enhance human analytic performance—particularly for anticipatory analyses or forecasts—emerged through a program of research, funded by the Intelligence Advanced Research Project Activity (IARPA), called *The Good Judgment Project* (GJP). The GJP was a large scale, multi-year research study examining human judgments, individually and collectively, about geopolitical affairs and events—a task closely related to intelligence analysis, at least for National Intelligence. They used a "forecasting tournament" approach, logging hundreds of thousands of forecasts from over two thousand participants. The participants were not all political "experts," but those who signed up to participate did have higher levels of intelligence and political knowledge than the general population (Mellers et al. 2015).

The GJP research team wanted to understand what distinguished more accurate from less accurate forecasters, whether forecasters did better working on their own or in groups, and how the performance of aggregated ("crowdsourced") forecasts might differ from individual ones. A particularly intriguing question was whether training in probabilistic thinking, scenario analysis, and tips for mitigating cognitive bias could improve the accuracy of forecasts. The prediction questions pertained to events such as who would win an upcoming Russian election and whether a violent incident would occur in the South China Sea in 2013 that kills at least one person (Mellers et al. 2015; Ungar et al. 2012).

Concerning the question of what dispositional and behavioral characteristics might distinguish more accurate from less accurate forecasters, they found:

> *The best forecasters scored higher on both intelligence and political knowledge than the already well-above-average group of forecasters. The best forecasters had more open-minded cognitive styles. They benefited from better working environments with probability training and collaborative teams. And while making predictions, they spent more time deliberating and updating their forecasts* (Mellers et al. 2015, p. 10).

Regarding the question of whether forecasters did better working on their own or in groups, and how the performance of aggregated forecasts might differ from individual ones, GJP researchers found that team forecasts (where groups of forecasters discussed the questions collectively deliberated on them) and prediction markets (e.g., aggregated forecasts) performed better than individual ones, but that forecast accuracy varied depending on exactly *how* the collective forecasts were combined.

Characteristics of the Best Geopolitical Forecasters:

- Open-minded
- Good at pattern detection
- High cognitive flexibility
- Aware of their own biases

(continued)

- Good at inductive reasoning
- Good understanding of geopolitics
- Understand probabilistic reasoning
- Reflective and intellectually curious
- Understand the nature of uncertainty
- Put forth sustained effort in deliberations
- Perform successfully in team environments
- Continuously monitor ongoing current affairs
- View forecasting as a skill, not an innate "talent"
- Good at disaggregating and synthesizing problem components
- Committed to practice, learning from, and improving decisions
- Adjust their forecasts more regularly, based on changing circumstances
- Skilled in balancing factors that affect judgments (e.g., over-under confidence)

Source: Mellers et al. (2015), Tetlock and Gardner (2016) and Ungar et al. (2012).

Finally, the GJP Researchers found that training did improve the accuracy of predictions. That may seem intuitively obvious, but quite a few prior studies have found that training in statistics, for example, did not mitigate the common probability biases or misleading heuristics that commonly plague human judgments. People tend to underestimate and have lower confidence about high probability events (above .75), and to overestimate with higher confidence those events with low probability (below .75) (Ferrell 1994). Statistical training and education have not eliminated those tendencies, but the GJP training did seem to improve forecasters' accuracy (Chang et al. 2016).

The training evolved over the years of the study. The nature of the training, and lessons for aspiring forecasters, are described elsewhere in greater detail (Chang et al. 2016; Tetlock and Gardner 2016), but by Year 3, they were focusing on a set of tips for better probabilistic reasoning and a set of guidelines for political reasoning. As a mnemonic device, they represented the main points by the acronym: CHAMPS KNOW, which breaks down in the following way (Chang et al. 2016; Tetlock and Gardner 2016):

Probabilistic Reasoning Principles

- **C**omparison classes inform probability estimates
- **H**unt for the right information
- **A**djust and update forecasts when appropriate
- **M**athematical and statistical models can help
- **P**ost-mortem analysis after outcomes are known
- **S**elect the right questions to answer

Political Reasoning Principles:

- **K**now the power players and their preferences
- **N**orms and protocols of domestic and international institutions matter
- **O**ther perspectives (bottom-up eruptions of mass protest, globalization/interconnected-world arguments, efficient-markets thesis and its critics …) should also inform your forecasts
- **W**ildcards, accidents and black swans can catch you off-guard. Beware of irreducible uncertainty

Results from the Good Judgment Project represent some of the most significant applications science to advance human performance on intelligence-related analytic assessments.

1.3 Technology-Centered Advances in Intelligence Analysis

In the early days of intelligence analysis, finding "enough" data and information was an ongoing challenge. Today, we have the opposite problem – information overload. The volume, variety, and velocity of information continues to grow with no clear leveling point in sight. The devices we carry and use every day to work, to communicate and to entertain are continuously generating and emitting new data from text and images to radio frequency signals to GPS coordinates. And with a surge in the Internet of Things (IoT), we are adding new data-generating devices every day (Brantly 2018).

We have encountered data surges before in modern history; for example, when active communication satellites first appeared in 1962, but what we face today is truly unprecedented. Putting a number on that volume of data is nearly impossible. The numbers used to estimate it are almost incomprehensible. Just when people have gotten their heads around the idea of gigabytes and terabytes, the amount of information extends to petabytes (each is a thousand terabytes) and zettabytes (each is a billion terabytes). To illustrate the volume and velocity of data, a 2014 report by Dell EMC estimated the size of the "digital universe" in 2013 to be 4.4 zettabytes. The report predicted by 2020, its size would grow tenfold to 44 zettabytes (Dell EMC 2014). Collecting, storing, managing, and analyzing that amount of data poses a whole new set of challenges to law enforcement and intelligence agencies (Akhgar et al. 2015; Chan and Moses 2017). As a result, data literacy, management, and analysis are rapidly emerging as core competencies for intelligence analysts (Hare and Coghill 2016; Manning 2008; Ratcliffe 2004; Weston et al. 2019).

While the continued rise of "Big Data" poses a number of challenges for law enforcement organizations, it also offers some compelling opportunities. As ILP's popularity continues to grow, the availability of more data and the ability to aggregate and analyze that data greatly enhance the potential for "predictive policing" and multi-dimensional crime control methods, which is likely to lead to more science-informed operational planning (Leng and Li 2018). The emergent field of "criminal analytics," based on Big Data, has important implications, not only for police patrols, but also for criminal investigation where tools may help to establish subject identities (resolving inconsistencies), reveal the structure of a criminal organization and the roles of its participants, link *modus operandi* across events, suggest subject location and profiles based on crime patterns and offense characteristics, and even shorten a list of suspects (Babuta 2017; Hu 2019; Li and Wang 2015; Perera et al. 2014; Pramanik et al. 2017).

In this section of the chapter, we will review some common and emergent technologies used by law enforcement intelligence professionals to exploit and derive insights from large amounts of data. Specifically, we will address spatial analytic technologies, network analytic technologies, data mining technologies, and the applications of artificial intelligence. These have been used in a wide range of policing applications, including human trafficking, wildlife trafficking, organized crime, and money laundering.

1.3.1 Spatial Analytic Technologies

Spatial analytic technologies have been a mainstay of crime analysis, almost since its inception. Spatial technologies, in the broadest sense, use place or location as the main unit of analysis. Before the advent of computer-assisted mechanisms, spatial analyses were manually represented on physical maps. It was an arduous, time-consuming process. These were the origins of "crime mapping." Early applications focused on the basic question of where, within a given jurisdiction, certain kinds of crimes were occurring. Because most crime in any given city is typically concentrated in 20% or less of the municipality's total area, this kind of mapping allowed police managers to allocate patrol resources where they were most needed.

Even when computerized crime mapping efforts first began to emerge in the late 1960s, however, the machinery was cumbersome and expensive and the results were very basic (akin to automated "pin mapping"). Crime mapping with Geographic Information Systems (GIS) did not really hit its stride until the late 1980s, as (a) information technologies advanced and became more affordable, (b) location-based crime reporting became mandatory, and (c) criminological approaches evolved from being offender-focused to more context-focused (Wang 2012).

The advent of the Global Positioning System (GPS) for geocoding markedly improved the accuracy and precision of GIS applications in law enforcement. In 1974, the US Government (USG) launched the first GPS satellite—part of its planned NAVSTAR System. The System currently has at least 32 operational satellites (with three "backups"), configured so that 24 are available 95% of the time. GPS was limited to military use until the early 1980s when it was opened to commercial civilian aircraft to improve safety and navigation. In 2000, Congress approved a plan to configure GPS satellites to transmit two additional signals that would be open for more widespread civilian use. Now, GPS receivers are inexpensive, widely available, and are even built into most smart phones. They are also commonly installed on police patrol vehicles and are increasingly attached to equipment that officers carry on their persons. This has allowed law enforcement agencies to optimize their dispatching and to more closely track their officers in the field.

Law enforcement use of spatial technologies has expanded more recently into the use of unmanned aerial vehicles (UAVs or "drones"). Civilian UAV use became more popular after Congress passed the *FAA Modernization and Reform Act of 2012* (FAA Act), which effectively cleared drone use for policing operations. In a 2015 study of 46 law enforcement agencies in 11countries, more than a quarter (26%) said they were using UAVs for policing-related purposes (Custers and Vergouw 2015). GPS receivers are also embedded in many advanced UAVs. Civilian police drones are often equipped with cameras and may be used for static (crowds, objects, crime scenes) or dynamic observation (moving vehicles, suspects). Some drones, such as those configured with emitting light, sound or scent, even have the potential for "active" use on the public (Engberts and Gillissen 2016). The potential use of "armed" drones, either with lethal or "less lethal" armaments, by domestic, civilian law enforcement agencies remains a matter of active debate and discussion (Brumfield 2014; Straub 2014).

Law enforcement GIS applications are still primarily used for crime prevention and increasingly for criminal investigations. GIS has had a central role in "predictive policing," a term used to describe the practice of using spatio-temporal, historical data to anticipate or predict future crimes (Ferreira et al. 2012; Hardyns and Rummens 2018; Herchenrader and Myhill-Jones 2015). Location data are used (a) to identify "hotspots," which may be small geographic areas with high crime concentrations or areas where risk zones are more densely concentrated around particular criminal incidents, and (b) to create Risk Terrain Models to identify areas that are sensitive to higher crime rates (Hardyns and Rummens 2018). A rigorous review of 19 experimental or quasi-experimental evaluation studies found that hotspot policing led to significant reductions in crime and/or disorder in 20 of the 25 instances in which it was tested (Braga et al. 2012). Hotspot policing is widely regarded as one of the most robustly supported evidence-based policing and crime policy practices. Beyond GIS' broader function in identifying risk zones and hotspots for crime, geographic profiling has also assisted investigators in individual cases by identifying an offender's most likely area of residence based on the locations of a linked series of crimes (Rossmo 2012; Rossmo and Velarde 2008).

1.3.2 Network Analytic Technologies

Understanding criminal networks is extraordinarily important for exploiting criminal enterprises, and even assessing the behavior of particular criminal actors (Bichler 2019). The yields can be especially valuable to intelligence analysts, particularly in cases involving clandestine activity, complex criminal enterprises, or operational cells (Burcher and Whelan 2018; Sparrow 1991). Clark (2019) suggests that ignoring or under-analyzing network activity is one of the major causes of intelligence failures. Network analyses—basically a map of the ties and interactions among of set of entities—have been successfully deployed in law enforcement against a wide range problems such as organized crime (van der Hulst 2009), drug trafficking (Bright et al. 2019), money laundering (Drezewski et al. 2015); terrorism (Dörfler et al. 2019); human trafficking, even trafficking wildlife (Haas and Ferreira 2015).

Different forms of network analysis exist. For example, basic Link Analysis and Analytic Charting have been staples in the analyst's toolkit for decades (Harper and Harris 1975). One of the most popular modern techniques among law enforcement and intelligence agencies has been Social Network Analysis (Van der Hulst 2009). About 43% of law enforcement agencies in one international sample said they employed SNA in their operations (Custers and Vergouw 2015). In SNA, people are typically the "entities" or "nodes" of interest. One key way in which SNA differs from other kinds of security-related analyses is that the focus is not on an individual offender, but on the ties and connections that each actor within a given system has or does not have with the other actors in that system. Simply put, relationships, not individual actors, are at the center of SNA (Van der Hulst 2009). Results are represented in visual form, which makes them easier for analysts to understand and to use in communicating network information to others (Xu and Chen 2005).

SNA reveals the structure of set of networked relationship. Analysts can interpret the implications of those structures for how the network will function and how its actors may behave (Custers and Vergouw 2015). Illicit and criminal networks do not always behave in the same way as legitimate and open networks (Bright et al. 2019). SNA can illuminate the roles and functions that actors may have in a network, and suggest important underlying dynamics like dependency between actors or relative power and position. It can visually represent and mathematically quantify the importance and centrality of the actors and the strength of ties between them (Bichler 2019; Jentner et al. 2018). SNA has been used successfully in law enforcement for scenario building, risk/threat assessments, hypothesis testing, network destabilization, and establishing identities in the presence of aliases and conflicting role information. Characterizing a criminal network in detail can even help intelligence planners in decisions about when, how, and where to deploy or insert human assets (Van der Hulst 2009).

1.3.3 Data Mining Technologies

Data mining is a broad term describing a range of tools and techniques for identifying patterns and extracting potential knowledge from large sets of data. Data mining techniques can be used for one or more of the following purposes: (1) to identify connections or associations between people or entities (Link Analysis, akin to SNA, which can be considered a Data Mining Technique); (2) to divide people or entities into clusters, classes, or categories; (3) to identify, define and describe concepts within a given knowledge domain; (4) to identify deviations or anomalies within data or data trends; and (5) to represent data visually to discern patterns and configurations that might not otherwise be detected in mathematical or numerical representations (Mishra and Shelke 2015).

These techniques have increasingly been applied in law enforcement contexts to support crime prevention, operational planning, and criminal investigations (McCue 2015; see Hassani et al. 2016 for an overview). Chen and colleagues (2003) provide the following examples of how various data mining techniques are used in policing:

> Entity extraction *has been used to automatically identify person, addresses, vehicles, narcotic drugs, and personal properties from police narrative reports.* Clustering techniques *such as "concept space" have been used to automatically associate different objects (such as persons, organizations, vehicles) in crime records.* Deviation detection *has been applied in fraud detection, network intrusion detection, and other crime analyses that involve tracing abnormal activities.* Classification *has been used to detect email spamming and find authors who send out unsolicited emails.* String comparators *have been used to detect deceptive information in criminal records.* Social network analysis *has been used to analyze criminals' roles and associations among entities in a criminal network" (pp. 1–2).*

There is substantial overlap between some data mining techniques and artificial intelligence (AI), primarily with regard to machine learning (ML). Machine learning is exactly what it sounds like. Algorithms are configured so that the software or "machine" actually learns as it sorts through the data. Two types of ML are commonly used, *supervised* and *unsupervised.* The difference pertains to whether the classification outcomes (or "ground truth") are known or unknown before the learning occurs.

For example, suppose you wanted to apply ML to distinguish fraudulent from legitimate account records. You could use supervised learning if you have a set of account records that you already know are fraudulent and a set that you know are not. The ML algorithms will examine all features or attributes of all accounts in the database (e.g., average balance, number of transactions in a given period, destinations of deposits, etc), compare those features between the two groups, "learn" which attributes differentiate those that are fraudulent from those that are legitimate, and build a model to predict the membership of new data points. If, however, you did not have a set of known fraudulent and legitimate accounts to start with, then the ML algorithms would need to sift through all attributes in all accounts to examine if two (or more) types of accounts could be identified based on the way the attributes cluster together or correlate with one another. If two (or more) clusters emerge from mining the data, then based on your prior knowledge (or what has been found in prior studies) and/or experience, you as the analyst would need to explore whether the way the attributes cluster and the differences between the two groups "make sense" or align with what might be expected in fraudulent and legitimate accounts. That would be an example of unsupervised learning. Although, in both cases, the software or machine learns from the data (learning is where the "intelligence" part of AI factors in), but what it learns, and how, is different.

Classification is one of the most fundamental and common uses for data mining and ML techniques. These techniques use rules or attributes within the data to sort entities into categories that are defined by those rules or attributes. In data mining, generally, there are three particularly common methods used for classification purposes: Decision Trees; Neural Networks; and Support Vector Machines. The following section discusses each of these in a bit more detail:

- **Decision Trees** use multistage algorithms (an algorithm is a formula that follows an established process or set of rules), ultimately leading to an endpoint outcome. The decision process is broken down into discrete segments, each with two or more options. Whichever of those options is selected or matched moves the formula to the next segment, that has two or more options. This process continues until the "answer" or ultimate outcome is reached. Decision Trees have been used to identify suspicious email messages and detect insurance fraud.
- **Artificial Neural Networks (ANNs)** comprise an interconnected set of component algorithms, modeled to function like the human brain, where each component (acting like a "processor") completes a small task that collectively contributes to the larger goal of recognizing a pattern. Typically, the network contains one or more "learning rules" that change the strength of the connection between components based on different inputs. ANNs do not process inputs in the logical, sequential way that typical serial computer processors do. The networks do not execute instructions; rather, they respond to the inputs based on the rules. These characteristics make ANNs especially useful when the relationships between variables are unknown or are complex and difficult to characterize. If the variables are *labeled*, the ANNs can classify them. If they are *unlabeled*, the ANNs can group or cluster them based on shared characteristics or attributes. "Through executing the learning process and using programmable memories, ANNs can predict new trends based on existing samples, and so they are being used to predict potential criminal patterns based on observations of current criminal activities" (Pramanik et al. 2017, pp. 5–6). ANNs have been applied to law enforcement tasks like detecting lies in a set of statements and identifying smuggling vessels.
- **Support Vector Machines (SVMs)** are often relatively simple algorithms used for supervised machine learning. Like all supervised learning, it begins with data from two known groups, from which it builds a model to predict group membership. Then the SVM sorts through the new data and classifies them into one of two groups or categories. The SVM's objective is to distinguish the two groups as clearly as possible, so that if they are plotted on a graph, the two clusters are visually as far apart from each other as possible. This can produce highly accurate results without requiring much computing power. SVMs can be used to categorize text, images, and even handwriting. SVMs have been applied in law enforcement tasks to identify harassing messages in social media posts, to identify the senders of emails and authors of documents, and for cybercrime detection.

1.3.4 Artificial Intelligence

Like Data Mining, the term Artificial Intelligence (AI) is used broadly to describe a wide range of computing technologies. A general definition might characterize AI as any software or hardware that either executes functions without human oversight or improves its own performance by learning from experience. The "intelligence" dimension refers to the ways in which AIs execute human-like tasks and functions, thinking and acting rationally in perceiving, acting, learning, deciding and solving new problems (Hoadley and Lucas 2018). For law enforcement applications, AI systems have the potential to engage in surveillance, intelligence collection and analysis, and even direct enforcement actions. A common example of AI-driven enforcement might occur when cameras capture a vehicle that enters the plane of an intersection after the traffic light turned red, sends a message to an information system indicating that a violation has occurred involving a vehicle with a given registration tag; that system links the tag number to the registrant, and automatically generates a citation with a designated fine, which is sent to the vehicle's registrant. When no human intervention is involved in the process, the AI's actions are often referred to as "automated law enforcement" (Petit 2018).

Despite extraordinary investments and improvements in AI technologies, particularly over the past 15 years, the ability of automated systems to surveil, collect and analyze intelligence, and take responsive enforcement action is currently limited. The leaps in progress may be most evident in some intelligence collection and analysis tasks, where AI systems have information gathering and processing capabilities that often exceed those of their human counterparts (Allen and Chan 2017; Ganor 2019). AI's computing advantage does not suggest that law enforcement officers and intelligence analysts can or should be replaced by machines. Nor is that kind of replacement currently feasible or even likely to occur within the foreseeable future (Faizal Bin Abdul Rahman 2017).

As with any technology trend, the rise of AI poses new opportunities and new challenges. The trend for AI applications in law enforcement lies in identifying ways that computerized or automated capabilities can support and improve human decision making (Eldridge et al. 2018; Llinas et al. 2017; Pike 2019). Intelligence analysis continues fundamentally to be a human task, and there are compelling reasons in law enforcement applications to keep human analysts at the center of the process. New spatial, network and data analytic technologies, however, can bring more data and different kinds of data to bear on critical decisions and strengthen the evidence base for analytic assessments. Algorithms and machine learning may be important components of that process (Regens 2019).

Analytic and AI technologies may be evolving to a point where they might effectively serve as consultants to investigators and analysts (Tecuci et al. 2014). "Intelligent agents" are a major focus area in AI research. The term does not refer to intelligence officers, assets or agents. The "agency" lies within the software or hardware itself. According to Brenner and colleagues (Brenner et al. 2012) "intelligent software agents are defined as being a software program that can perform specific tasks for a user and possesses a degree of intelligence that permits it to perform parts of its tasks autonomously and to interact with its environment in a useful manner" (p. 21). These agents can identify, extract and monitor information for analysis and potentially for response (Brenner et al. 2012). This kind of automated action could provide invaluable support for law enforcement intelligence analysts.

Although the abilities of AI technologies to classify and predict are constantly improving, data mining and AI technologies are quite limited when it comes to "understanding" phenomena and determining causal relationships. They may predict an outcome with reasonable accuracy, but they cannot explain why that decision was reached. While AI and data analytics can enable and accelerate LE analytic capabilities, humans infuse meaning and sense making into the results (Lim 2016). Data mining and AI may best support human analysts by triaging large volumes of data and extracting key elements to hone the analyst's focus (Llinas et al. 2017). Of course, to properly leverage (and contain) these technologies, proficiencies in data science and technology are likely to become increasingly important skills for law enforcement intelligence analysts (Eldridge et al. 2018).

2 Conclusions

Intelligence-led policing has become a popular, in some cases even dominant, approach to managing law enforcement operations in many Western countries (Bureau of Justice Assistance 2005; Ratcliffe 2016). This philosophical shift has elevated the importance of intelligence analytic capabilities in police organizations, and put intelligence analysts at the center of critical operational decisions (Burcher and Whelan 2019). Over the past 20 years, trends in professionalism and professional standards within the field of law enforcement intelligence analysis, and external developments in science and technology have considerably advanced analytic practice.

Looking at the human side of scientific advances, the emergence of two major professional organizations—IALEIA and IACA—have moved law enforcement analysts toward a sense of collective identity and created an infrastructure to improve both analyst qualifications and analytic products. Although each organization has a slightly different analytic focus (traditional vs data-driven analytic tradecraft) both offer analytic training, have the reference sources for a common body of knowledge, and have established professional certification programs. Also, cognitive psychologists have been giving greater attention to researching analytic tasks, and intelligence analysts have been following that research more closely and applying it to their own tradecraft. A very recent scientific research program—the Good Judgment Project—has begun to illuminate "what works" in forecasting future (particularly geopolitical) events, and the kind of traits and abilities that characterize people who consistently make the most accurate predictions. This cross-pollination of science and practice may lead to greater reliance on science-informed intelligence analysis in policing.

On the technological side of advances in the field, a range of data analytic technologies have emerged to help analysts manage the influx of Big Data. Spatial technologies can identify location patterns among people, objects or events. Network analyses can illuminate how people, objects or events are connected or tied together. Data Mining technologies can find patterns and anomalies amidst massive volumes of structured or unstructured data. And Artificial Intelligence technologies can learn from sifting through datasets; can cluster, classify or categorize the data; and can identify the more reliable and relevant data points, reducing the data load for analysts. Machine-driven analytic technologies will continue to improve, and will likely do so at an incredible pace. Their applications in law enforcement operations will test legal, ethical and policy boundaries, but they cannot and will not replace critical capabilities provided by trained human analysts.

References

Ackoff, R. L. (1989). From data to wisdom. *Journal of Applied Systems Analysis, 16*(1), 3–9.

Akhgar, B., Saathoff, G. B., Arabnia, H. R., Hill, R., Staniforth, A., & Bayerl, P. S. (2015). *Application of big data for national security: A practitioner's guide to emerging technologies*. Waltham: Butterworth-Heinemann.

Allen, G., & Chan, T. (2017). *Artificial intelligence and national security* (Vol. 132). Cambridge, MA: Belfer Center for Science and International Affairs.

Artner, S., Girven, R. S., & Bruce, J. B. (2016). *Assessing the value of structured analytic techniques in the US intelligence community* (No. RR-1408-OSD). Santa Monica: RAND National Defense Research Institute.

Babuta, A. (2017). Big data and policing: An assessment of law requirements, expectations and priorities. In *RUSI occasional paper*. London: Royal United Services Institute for Defence and Security Studies. Retrieved from https://rusi.org/sites/default/files/201709_rusi_big_data_and_policing_babuta_web.pdf

Bichler, G. (2019). *Understanding criminal networks: A research guide*. Oakland: University of California Press.

Braga, A., Papachristos, A., & Hureau, D. (2012). Hot spots policing effects on crime. *Campbell Systematic Reviews, 8*(8), 1–96.

Brantly, A. F. (2018). When everything becomes intelligence: Machine learning and the connected world. *Intelligence and National Security, 33*(4), 562–573. https://doi.org/10.1080/02684527.2018.1452555.

Brenner, W., Zarnekow, R., & Wittig, H. (2012). *Intelligent software agents: Foundations and applications*. New York: Springer Science & Business Media.

Bright, D., Koskinen, J., & Malm, A. (2019). Illicit network dynamics: The formation and evolution of a drug trafficking network. *Journal of Quantitative Criminology, 35*(2), 237–258.

Bruce, J. B., & George, R. (2015). Professionalizing intelligence analysis. *Journal of Strategic Security, 8*(3), 1–23.

Brumfield, E. (2014). Armed drones for law enforcement: Why it might be time to re-examine the current use of force standard. *McGeorge Law Review, 46*, 543–572.

Burcher, M., & Whelan, C. (2018). Social network analysis as a tool for criminal intelligence: Understanding its potential from the perspectives of intelligence analysts. *Trends in Organized Crime, 21*(3), 278–294. https://doi.org/10.1007/s12117-017-9313-8.

Burcher, M., & Whelan, C. (2019). Intelligence-led policing in practice: Reflections from intelligence analysts. *Police Quarterly, 22*(2), 139–160. https://doi.org/10.1177/1098611118796890.

Bureau of Justice Assistance. (2005, September). *Intelligence-led policing: The new intelligence architecture*. Washington, DC: Bureau of Justice Assistance. Retrieved from: https://www.ncjrs.gov/pdffiles1/bja/210681.pdf

Carter, D. L. (2009). *Law enforcement intelligence: A guide for state, local, and tribal law enforcement agencies*. Washington, DC: US Department of Justice, Office of Community Oriented Policing Services.

Carter, D. L., & Carter, J. G. (2009). Intelligence-led policing: Conceptual considerations for public policy. *Criminal Justice Policy Review, 20*(3), 310.

Chan, J., & Moses, L. B. (2017). Making sense of big data for security. *British Journal of Criminology, 57*(2), 299–319. https://doi.org/10.1093/bjc/azw059.

Chang, W., Chen, E., Mellers, B., & Tetlock, P. (2016). Developing expert political judgment: The impact of training and practice on judgmental accuracy in geopolitical forecasting tournaments. *Judgment and Decision making, 11*(5), 509–526.

Chen, H., Chung, W., Qin, Y., Chau, M., Xu, J. J., Wang, G., Zheng, R., & Atabakhsh, H. (2003, May). Crime data mining: an overview and case studies. In *Proceedings of the 2003 annual national conference on Digital government research* (pp. 1–5). Digital Government Society of North America.

Clark, R. M. (2019). *Intelligence analysis: A target-centric approach (6th Ed.)*. Washington, DC: CQ Press.

Custers, B., & Vergouw, B. (2015). Promising policing technologies: Experiences, obstacles and police needs regarding law enforcement technologies. *Computer Law & Security Review, 31*(4), 518–526.

Dell EMC. (2014). *The digital universe of opportunities: Rich data and the increasing value of the internet of things*. Hopkinton: Dell EMC.

Dörfler, T., Stollenwerk, E., & Schibberges, J. (2019). *Uncovering covert structures: Social network analysis and terrorist organizations*. London: SAGE.

Drezewski, R., Sepielak, J., & Filipkowski, W. (2015). The application of social network analysis algorithms in a system supporting money laundering detection. *Information Sciences, 295*, 18–32. https://doi.org/10.1016/j.ins.2014.10.015.

Eldridge, C., Hobbs, C., & Moran, M. (2018). Fusing algorithms and analysts: Open-source intelligence in the age of 'Big Data'. *Intelligence and National Security, 33*(3), 391–406. https://doi.org/10.1080/02684527.2017.1406677.

Engberts, B., & Gillissen, E. (2016). Policing from above: Drone use by the police. In B. Custers (Ed.), *The future of drone use* (pp. 93–113). The Hague: TMC Asser Press.

Faizal Bin Abdul Rahman, M. (2017). Why it won't displace police analysts artificial intelligence. *RSIS Commentary, 109*, 1–3.

Ferreira, J., João, P., & Martins, J. (2012). GIS for crime analysis: Geography for predictive models. *Electronic Journal of Information Systems Evaluation, 15*(1), 36.

Ferrell, W. R. (1994). Discrete subjective probabilities and decision analysis: Elicitation, calibration and combination. In G. Wright & P. Ayton (Eds.), *Subjective probability* (pp. 411–451). Chichester: Wiley.

Ganor, B. (2019). Artificial or human: A new era of counterterrorism intelligence? *Studies in Conflict and Terrorism, 0*(0), 1–20. https://doi.org/10.1080/1057610X.2019.1568815.

Gentry, J. A. (2016). The "professionalization" of intelligence analysis: A skeptical perspective. *International Journal of Intelligence and CounterIntelligence, 29*(4), 643–676.

Global Justice Information Sharing Initiative. (2003). *The national criminal intelligence sharing plan*. Washington, DC: Global Justice Information Sharing Initiative, U.S. Department of Justice.

Global Justice Information Sharing Initiative. (2007). *Law enforcement analyst certification standards*. Washington, DC: Global Justice Information Sharing Initiative, U.S. Department of Justice.

Haas, T. C., & Ferreira, S. M. (2015). Federated databases and actionable intelligence: Using social network analysis to disrupt transnational wildlife trafficking criminal networks. *Security Informatics, 4*(1), 1–14. https://doi.org/10.1186/s13388-015-0018-8.

Hardyns, W., & Rummens, A. (2018). Predictive policing as a new tool for law enforcement? Recent developments and challenges. *European Journal on Criminal Policy and Research, 24*(3), 201–218. https://doi.org/10.1007/s10610-017-9361-2.

Hare, N., & Coghill, P. (2016). The future of the intelligence analysis task. *Intelligence and National Security, 31*(6), 858–870. https://doi.org/10.1080/02684527.2015.1115238.

Harper, W. R., & Harris, D. H. (1975). The application of link analysis to police intelligence. *Human Factors, 17*(2), 157–164.

Harrison, M., Walsh, P. F., Lysons-Smith, S., Truong, D., Horan, C., & Jabbour, R. (2018). Tradecraft to standards—Moving criminal intelligence practice to a profession through the development of a criminal intelligence training and development continuum. *Policing: A Journal of Policy and Practice*, 1–13. https://doi.org/10.1093/police/pay053.

Hassani, H., Huang, X., Silva, E. S., & Ghodsi, M. (2016). A review of data mining applications in crime. *Statistical Analysis and Data Mining: The ASA Data Science Journal, 9*(3), 139–154.

Herchenrader, T., & Myhill-Jones, S. (2015). GIS supporting intelligence-led policing. *Police Practice and Research, 16*(2), 136–147.

Heuer, R. J. (1999). *Psychology of intelligence analysis*. Washington, DC: Center for the Study of Intelligence, Central Intelligence Agency.

Heuer, R. J., & Pherson, R. H. (2010). *Structured analytic techniques for intelligence analysis*. Washington, DC: CQ Press.

Hoadley, D. S., & Lucas, N. J. (2018). *Artificial intelligence and national security*. Washington, DC: Congressional Research Service.

Hu, J. (2019). Big data analysis of criminal investigations. In *2018 5th international conference on systems and informatics, ICSAI 2018* (pp. 649–654). https://doi.org/10.1109/ICSAI.2018.8599305.

Jentner, W., Sacha, D., Stoffel, F., Ellis, G., Zhang, L., & Keim, D. A. (2018). Making machine intelligence less scary for criminal analysts: Reflections on designing a visual comparative case analysis tool. *The Visual Computer, 34*(9), 1225–1241. https://doi.org/10.1007/s00371-018-1483-0.

Leng, J., & Li, G. (2018). Big data-driven predictive policing innovation. *Advances in Engineering Research, 163*, 123–127. https://doi.org/10.2991/iceesd-18.2018.19.

Li, J., & Wang, A. G. (2015). A framework of identity resolution: Evaluating identity attributes and matching algorithms. *Security Informatics, 4*(1). https://doi.org/10.1186/s13388-015-0021-0.

Lim, K. (2016). Big data and strategic intelligence. *Intelligence and National Security, 31*(4), 619–635.

Llinas, J., Rogova, G., Barry, K., Hingst, R., Gerken, P., & Ruvinsky, A. (2017). Reexamining computational support for intelligence analysis: A functional design for a future capability. *Autonomy and Artificial Intelligence: A Threat or Savior?* (April 2018), 13–46. https://doi.org/10.1007/978-3-319-59719-5_2

Manning, P. (2008). *The technology of policing: Crime mapping, information technology, and the rationality of crime control*. New York: New York University Press.

Marrin, S. (2016). Improving intelligence studies as an academic discipline. *Intelligence and National Security, 31*(2), 266–279.

Marrin, S., & Clemente, J. D. (2006). Modeling an intelligence analysis profession on medicine. *International Journal of Intelligence and CounterIntelligence, 19*(4), 642–665.

McCue, C. (2015). *Data mining and predictive analysis: Intelligence gathering and crime analysis*. Oxford/Waltham: Butterworth-Heinemann.

Mellers, B., Stone, E., Atanasov, P., Rohrbaugh, N., Metz, S. E., Ungar, L., Bishop, M., Horowitz, M., Merkle, E., & Tetlock, P. (2015). The psychology of intelligence analysis: Drivers of prediction accuracy in world politics. *Journal of Experimental Psychology: Applied, 21*(1), 1–14.

Mishra, N., & Shelke, P. (2015). Data mining – A necessity for crime detection. *International Journal on Recent and Innovation Trends in Computing and Communication, 3*(2), 291–294. Retrieved from http://www.ijritcc.org

Perera, H., Udeshini, S., & Munasinghe, M. (2014, November). *Criminal short listing and crime forecasting based on modus criminal short listing and crime*. https://doi.org/10.13140/RG.2.1.5149.1287

Petit, N. (2018). Artificial intelligence and automated law enforcement: A review paper. *SSRN Electronic Journal, 1*(1974), 1–12. https://doi.org/10.2139/ssrn.3145133.

Pike, T. (2019, May). *Computational tools to support analysis and decision making*, 23. https://doi.org/10.1117/12.2518693.

Pramanik, M. I., Lau, R. Y. K., Yue, W. T., Ye, Y., & Li, C. (2017). Big data analytics for security and criminal investigations. *Wiley Interdisciplinary Reviews: Data Mining and Knowledge Discovery, 7*(4), 1–19. https://doi.org/10.1002/widm.1208.

Ratcliffe, J. H. (2004). Crime mapping and the training needs of law enforcement. *European Journal on Criminal Policy and Research, 10*(1), 65–83.

Ratcliffe, J. (2016). *Intelligence-led policing*. New York: Routledge.

Regens, J. L. (2019). Augmenting human cognition to enhance strategic, operational, and tactical intelligence. *Intelligence and National Security, 34*(5), 673–687. https://doi.org/10.1080/02684527.2019.1579410.

Rossmo, D. K. (2012). Recent developments in geographic profiling. *Policing: A Journal of Policy and Practice, 6*(2), 144–150.

Rossmo, D. K., & Velarde, L. (2008). Geographic profiling analysis: Principles, methods and applications. In S. Chainey & L. Thompson (Eds.), *Crime mapping case studies: Practice and research* (pp. 35–43). New York: Wiley.

Rowley, J. (2007). The wisdom hierarchy: Representations of the DIKW hierarchy. *Journal of Information Science, 33*(2), 163–180.

Sparrow, M. K. (1991). The application of network analysis to criminal intelligence: An assessment of the prospects. *Social Networks, 13*(3), 251–274.

Straub, J. (2014). Unmanned aerial systems: Consideration of the use of force for law enforcement applications. *Technology in Society, 39*, 100–109.

Tecuci, G., Schum, D. A., Marcu, D., & Boicu, M. (2014). Computational approach and cognitive assistant for evidence-based reasoning in intelligence analysis. *International Journal of Intelligent Defence Support Systems, 5*(2), 146. https://doi.org/10.1504/ijidss.2014.059976.

Tetlock, P. E., & Gardner, D. (2016). *Superforecasting: The art and science of prediction*. London: Random House.

Ungar, L., Mellers, B., Satopää, V., Tetlock, P., & Baron, J. (2012, October). *The good judgment project: A large scale test of different methods of combining expert predictions*. AAAI technical report FS-12-06: Machine Aggregation of Human Judgment, 37–42.

Van der Hulst, R. C. (2009). Introduction to Social Network Analysis (SNA) as an investigative tool. *Trends in Organized Crime, 12*(2), 101–121. https://doi.org/10.1007/s12117-008-9057-6.

Wang, F. (2012). Why police and policing need GIS: An overview. *Annals of GIS, 18*(3), 159–171.

Warner, M. (2002). Wanted: A definition of "intelligence". *Studies in Intelligence, 46*(3), 15–22.

Weston, C., Bennett-Moses, L., & Sanders, C. (2019). The changing role of the law enforcement analyst: Clarifying core competencies for analysts and supervisors through empirical research. *Policing and Society*. https://doi.org/10.1080/10439463.2018.1564751.

Xu, J., & Chen, H. (2005). Criminal network analysis and visualization. *Communications of the ACM, 48*(6), 100–107.

Improving Criminal Investigations with Structured Analytic Techniques

Max M. Houck

Abstract The intelligence community has used structured analytic techniques, methods designed to reduce bias and increase transparency of process, for years. The techniques force analysts out of routine thinking and away from heuristic habits in order to increase creativity, more comprehensively evaluate the questions, and create a document trail that reveals the thinking process that led to the intelligence product. These methods can be adapted for use in criminal investigations to help reduce bias, improve accuracy, and avoid both wrongful convictions (over 2200 to date) and reparations (more than $2.2 billion) while optimizing resources. The methods shift the investigator from intuitive, daily thinking (System 1, in Kahneman's terminology) to a more analytical approach (System 2) that creates a transparent process, regardless of the outcome. Structured analytical techniques are simple to use, inexpensive, and largely visual; they promote transparency, creativity, and group discussion, leading to better-supported results.

Keywords Criminal investigations · Structured analytic techniques · Intelligence

Doubt is not a pleasant state, but certainty is a ridiculous one. –Voltaire

If the only tool you have is a hammer, it is tempting to treat everything as if it were a nail. —Abraham Maslow

1 Introduction

For the last 100 years, criminal investigations have increasingly relied upon evidence beyond mere eyewitness information. The majority of this evidence is forensic in nature, that is, requiring scientific expertise to collect, analyze, and interpret.

M. M. Houck (✉)
Forensic Studies & Justice Program, University of South Florida, St. Petersburg, FL, USA
e-mail: maxmhouck@mail.usf.edu

B. Fox et al. (eds.), *Science Informed Policing*, Advanced Sciences and Technologies for Security Applications,
https://doi.org/10.1007/978-3-030-41287-6_7

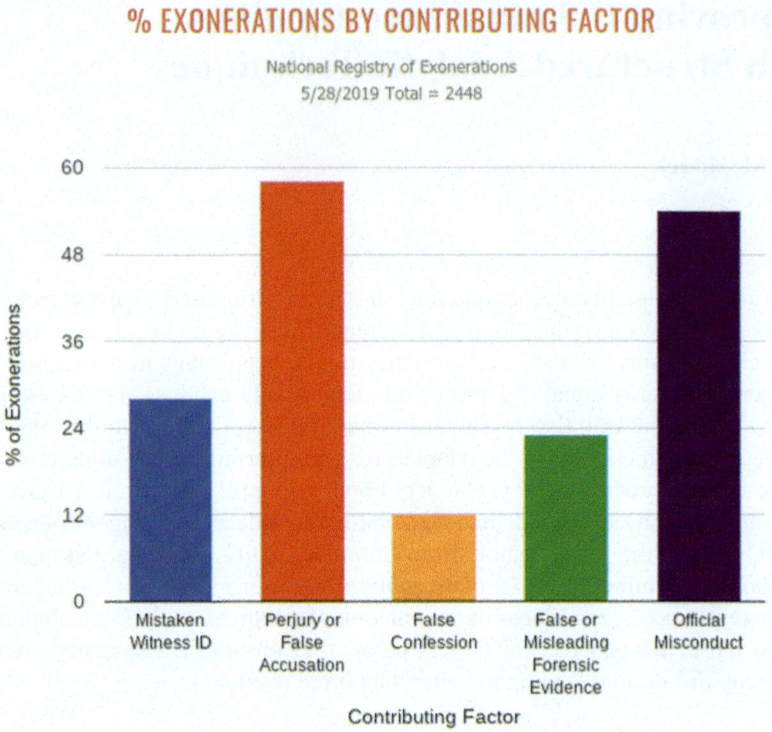

Fig. 1 Percent exonerations by contributing factor, as of 28 May 2019. (Source: National Registry of Exonerations)

A forensic investigation, then, is an inquiry to reliably address case-related questions about the evidence, its relevance in context, and ultimately meet the challenges of courtroom admissibility (if required to). Forensic investigations involve information gathering (intelligence), crime scene investigation, laboratory analysis, interpretation of tests and results, and reporting and communicating. For nearly every type of evidence, be it chemical, biological, or physical, the ability to detect less and less and produce more results has improved the potential to correctly solve cases and promote justice.

Police work and investigations have been largely procedural and intuitive. This makes them subject to a variety of biases, both implicit and cognitive (Spencer et al. 2016; Smith and Alpert 2007) that lead to poor policing, wrongful arrests and convictions, and unnecessary incarcerations. According to the National Registry of Exonerations,[1] almost 2500 wrongfully convicted citizens have been identified and exonerated (Fig. 1). Almost 22,000 years of those peoples' lives were lost to wrongful incarceration. The toll on their lives, their families, their friends, and society is morally indefensible. And those are just the identified: Estimates of wrongfully

[1] http://www.law.umich.edu/special/exoneration/Pages/about.aspx

incarcerated individuals run as high as 4% (Gross et al. 2014). Some jurisdictions have paid reparations to the exonerated, totalling over $2.2 billion (National Registry of Exonerations). Those reparations are cold comfort but consider what could have been achieved if those funds had gone to improved investigations and preventative measures. Public money in support the fair application of justice results in improved security and builds goodwill between the citizenry and law enforcement.

This chapter outlines the adaptation and integration of methods used for years in the intelligence community to criminal investigations. The methods are designed to improve accuracy, transparency, and promote documented procedures to facilitate audits and reviews.

2 Why Structured Analytic Techniques?

The intelligence community (or IC for short) has been using structured analytic techniques (SATs) for some time. The awareness of the need for SATs was made clear by a book written by Richards Heuer, a 45-year CIA analyst and researcher. The book, *Psychology of Intelligence Analysis* (Heuer 2015), lays out the basic tenets of analysis:

1. Humans are inherently bad at dealing with uncertainty
2. Awareness of this flaw does not help overcome it
3. Techniques and methods that make reasoning explicit can mitigate this flaw

Heuer's book emphasized the benefits of using techniques collaboratively to make the analytic process explicit, documented, and open to questioning, thus improving the chances for accuracy and completeness. Heuer's work was expanded and detailed in a book co-authored with Randolph Pherson, *Structured Analytic Techniques for Intelligence Analysis* (Heuer and Pherson 2015). The SATs have been demonstrated to improve intelligence analysis (Artner et al. 2016).

The purpose of this chapter is to apply SATs to investigations, especially those that involve forensic evidence. The IC and investigative professions share many similarities in how they approach their subject matter. Both must overcome cognitive biases and inherent or induced uncertainties that dog their methods and outcomes:

- Limited or incomplete information
- Unreliable, conflicting, or ambiguous information
- Denial and deception
- Information in the context of volatile or unknowable social situations
- Work within limited time frames
- Collect appropriate information
- Identify information gaps

For both the IC and criminal investigators, it is vitally important to get the best answer possible.

Why should criminal investigators use intelligence methods? In short, the growing complexity of and multidisciplinary input to many if not most investigations. Modern investigations typically involve multiple investigators, emphasizing the need for demonstrating information and documenting processes to share across agency boundaries. Expertise is distributed across individuals and these professional boundaries may become fuzzy, leading to miscommunication or mistakes. The diversity of incidents and the potential for significant mistakes highlights the need to identify alternative mental models and evaluate their validity (George and Bruce 2014).

Investigations of all types deal with incomplete, ambiguous, and deceptive information. SATs externalize internal thought processes systematically and transparently. The work product of SATs can be shared, edited, and critiqued collaboratively so that differing viewpoints or arguments are voiced early in the investigative process and open to revision in the light of significant information. SATs provide a solid analytic framework that documents how decisions were made and why (Heuer 2015). As Heuer notes (2015, page 10),

> Analysts do not achieve objective analysis by avoiding preconceptions; that would be ignorance or self-delusion. Objectivity is achieved by making basic assumptions and reasoning as explicit as possible so that they can be challenged by others and analysts can, themselves, examine their validity.

Citing the rationale for developing and employing SATs, Heuer (2015) notes that people tend to perceive what they expect (or want) to perceive. These expected perceptions quickly form outlooks or mindsets that are resistant to change, even assimilating new information into existing viewpoints to reduce cognitive dissonance. Ambiguous or vague information clouds or dilutes a more accurate perception and is either incorporated and skewed to support the preconceived ideas or discarded as irrelevant or unsupported.

The two main goals of SATs are to make reasoning transparent and document the process (Heuer and Pherson 2015). Because biases and flawed thinking are inherent to the human reasoning process, it is not a matter of *whether* prior assumptions and expectations will influence analysis—they will. The best way to control this influence is to make the process transparent and create a permanent record available for sharing and review.

2.1 System 1 and System 2

The Nobel Prize-winning work of Daniel Kahneman and Amos Tversky has led to a better understanding of how people think and reason—or don't. The human consciousness is divided into two systems, System 1 and System 2 (Kahneman 2011). System 1 is fast, automatic, emotional, subconscious, primitive, and runs automatically (and cannot be turned off). It is intuitive and powerful but also impatient and impulsive. System 1 can work on many issues or levels at the same time while using little energy. It quickly creates meaning out of things and is easily influenced: It is as likely to produce an apophany[2] as an epiphany. System 2, by contrast, is slow, effortful, logical, calculating, and must be consciously engaged. It is methodical, cautious, has limited capacity, and is singularly focused. System 2 is slow and a heavy energy user. Also, System 2 is very good at making comparisons but does not perform well under pressure. Under normal conditions, System 2 can override System 1 but if System 1 is stimulated (high emotions or stress, for example) or System 2 is tired or preoccupied, then System 2 may fail to control System 1. For example, a class of students taking a math test are all making System 2 decisions but their levels of right answers will vary enormously because of knowledge retention, stress, and other circumstances. If the students' information is bad, their answers will be bad, also. Engaging System 2 does not ensure an accurate outcome, just a more analytic one. People can engage System 1 in what *feels* like an analytic process when it is not. For example, if a baseball bat and ball cost $1.10 and the bat costs one dollar more than the ball, how much does the ball cost? The correct answer is often surprising, confusing, and may require a bit of convincing.[3]

2.1.1 Why System 1 Can Be Right…Sometimes

Human minds are ill-equipped to cope effectively with both inherent and induced uncertainty. Two million or more years of human evolution have produced a system that is quick to judge for survival's sake. Rustling in the grass is more likely a predator than a friend and running is the most expedient response; if it was not a predator, then nothing lost and a funny story gained. System 1 has been a constant with humans and was, for millions of years, key to our survival in the wild. But System 1 can also work in our favor in the modern world. If you were asked which city has the larger population, Munich or Dusseldorf, what be your response? You don't really know the populations, do you? But, like most people, you would guess Munich because you have heard that city's name more than you have Dusseldorf, so you go with Munich. Turns out, you would be right: Munich has about 1.3 million people and Dusseldorf has about half that (594,000).

This is the Recognition Heuristic (Goldstein and Gigerenzer 2002). Heuristics, essentially shortcuts in thinking, allow people to be mostly right about most things and have served an adaptive evolutionary purpose, like knowing to run from rustling

[2] The tendency to mistakenly perceive connections and meaning between unrelated things. A classic example of apophany is Skinner's Box, where a hungry pigeon was put in a box and randomly given food pellets. Because the pigeon got a food pellet while performing some action, it correlated the food and the action. As the pigeon increased the number of times it performed that action, it strengthened the false correlation between the non-correlated events (the feeding was random) (Skinner 1948). In humans, one example is the Gambler's Fallacy, where a perceived "hot streak" is in fact related to chance (Burns and Corpus 2004).

[3] If your answer was $0.10, that's System 1 talking. If the bat costs $1 <u>more than</u> the ball, the ball has to be $0.05 ($1.05 + $0.05 = $1.10). If the ball cost $0.10, the bat *and* ball would cost $1.20.

grass. The human brain makes mental models (frames) to deal with the uncertainty in complex situations, like criminal investigations. Mental models are necessary for people to process what otherwise would be a bewildering torrent of information.

But System 1 is easily foiled or high-jacked. Humans rely on heuristics but they are heavily influenced by what can be immediately recalled or what is directly in front of us. Emotions, perception, and context can easily derail analytical decision making. System 2 can also be influenced by System 1. In one study, it was suggested to fingerprint experts that a set of prints to be examined would not match, not knowing they had previously positively associated the prints in casework. Four out of five of the experts contradicted their previous conclusions (the prints were not a match or were inconclusive) (Dror et al. 2006). A similar phenomenon has been found in other forensic analyses, including DNA (Dror and Hampikian 2011).

2.2 *The Incompetent and the Overconfident*

Professionals are most dangerous at two points in their career, the beginning and the end. At the beginning they are "unconsciously incompetent," that is, they are not aware of what they do not know (the "unknown unknowns" of a Johari window, discussed later). Experienced experts, however, suffer from "unconscious competence" in that they know what they know but may have built all or part of that expertise on faulty experience; it is difficult for them to unpack what they know and explain how they know it. The specifics of both states are discussed below.

2.2.1 The Incompetent

McArthur Wheeler robbed two banks in Pittsburgh, PA. He was not wearing a mask and his face was clearly seen on surveillance cameras: Wheeler was caught on tape, in the bank, with a gun in his hand. The news broadcast his image; less than an hour after the broadcast, he was arrested. When he was arrested, Wheeler expressed complete disbelief at his identification and capture. He complained to the arresting officer, "But I wore the juice." He thought that rubbing lemon juice on his face would make him invisible to the cameras, much like lemon juice is used as "invisible ink."[4] Wheeler had even practiced at home with an instant camera and his face was not in any of the photographs. Detectives figured that the film was bad, the camera was adjusted incorrectly, or Wheeler had pointed the camera away from his face when

[4] Carbon-based compounds in the juice are absorbed into the paper's fibers. Because lemon juice ink is a weak acid, it softens the fibers in the paper. When the paper is warmed, the added heat causes some of the chemical bonds in the dried juice to break down and carbon is released. When the carbon comes into contact with air, it oxidizes, turning the juice pattern brown and, thus, visible.

he took the picture. Detailed in their landmark 1999 paper, Kruger and Dunning note (page 1121),

> When people are incompetent in the strategies they adopt to achieve success and satisfaction, they suffer a dual burden: Not only do they reach erroneous conclusions and make unfortunate choices, but their incompetence robs them of the ability to realize it. Instead, like Mr. Wheeler, they are left with the erroneous impression they are doing just fine.

Wheeler's incompetence was applied to illegal activities, but even with the best of intentions, people's incompetence prevents them from realizing that they are incompetent: They do not know what they do not know. This is called the Dunning-Kruger Effect.

2.2.2 The Overconfident

Confidence is a poor predictor of accuracy. Much research has established that exposure to or experience with a particular method or activity does not make one an expert and this is true across a wide range of professions (Ericsson et al. 2018). Simply making more decisions in a domain increases one's confidence, regardless of the accuracy of those decisions. A person's confidence in a decision increases over time (confidence hardening) and this is particularly true if they think they have received confirmation of that choice. Experts are not immune from this phenomenon. A stunning example of this is research done on the accuracy of facial comparisons done by Australian passport office staff (White et al. 2014). Staff were asked to conduct facial comparison and identification—a task the do daily—and the results were compared to their years in that position. The individual differences were very large and average performance was poor. While some staff were very accurate, this accuracy did not correlate with length of experience or training (Fig. 2).

More information is often believed to improve the decision process. In reality, the outcome of more information is reduced accuracy, creating an "illusion of knowledge" (Hall et al. 2007). A number of factors lead to this reduced accuracy. When a person has more information about a situation or topic, it means they have more awareness of variance and, therefore, they produce more "scatter" in their thinking (more possibilities and nuances) and less consistent judgments. Second, the added information also may be of a low quality, diluting the diagnosticity of the earlier information. Finally, the test data used to create or form a decision process fits it well but "wild" data does not (in statistics, this is called "overfitting") (Nisbett et al. 1981). In one example, a group of gamblers were given information about horses to bet on in a series of races. With only five items of information, the gamblers' confidence was well calibrated with their accuracy, but they became more confident as additional information was received without any increased accuracy (Slovic 2018) (Fig. 3).

The implications in an investigative analysis are significant: Given a biased viewpoint (the suspect is guilty), as more information is accrued and interpreted as only confirming that viewpoint, the investigator's confidence increases but does not

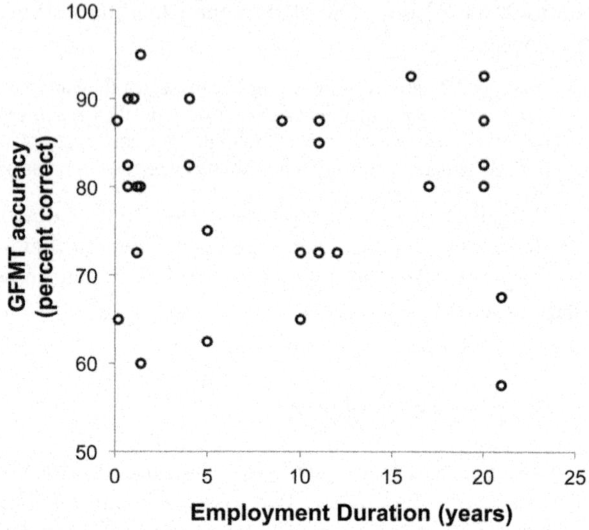

Fig. 2 Comparison of facial comparison accuracy with years of experience in that task. Some staff performed with near-perfect accuracy and some performed close to chance. Accuracy did not correlate with years employed in that task. Experience does not equal expertise. (Source: White et al. 2014)

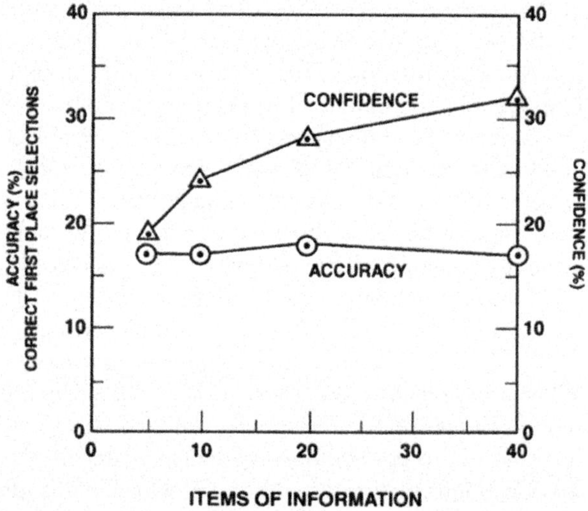

Fig. 3 Gamblers' confidence in horse racing bets increased with more information; the additional information did not make them more accurate, however. (Data from Slovic 2018; plot from Heuer 2015)

improve the accuracy of that position (the suspect actually may be innocent). And, accurate or not, making more decisions within a domain always leads to more confidence in those decisions (Einhorn and Hogarth 1978). As the saying goes, the plural of "anecdote" is not "data." An analyst's personal experience may be a poor guide to the revision of his or her mental model: A person can make the same mistake for years and call it experience.

2.2.3 Cognitive Biases

A cognitive bias refers to a systematic pattern of deviation from rationality in judgment. The difficulty is cognitive biases happen mostly unconsciously. Becoming more aware of biases can help somewhat but is by no means a complete response. Cognitive biases are a bit like optical illusions of thinking: Their effects require effort to work through (Fig. 4).

Cognitive biases are not optical, however, and they often prevent a clear path to an accurate answer. For example, take the following situation:

> Two men were in a bar. Two women walked in. The first man said, "I have to go, my wife and daughter are here." The second man turned around and said, "I have to go, also. My wife and daughter just arrived as well."

How is this possible? Our preconceived notions, implicit biases, and cultural expectations impede what is a simple and almost obvious answer (see the References section for the answer, but think about it before you do). Biases are closely linked with System 1; awareness of the bias alone is not enough to overcome it. Structured analytic techniques are inherently *debiasing* techniques: They do not replace intuitive judgment, they augment it and make it more resistant to bias. No formula exists for always getting it right but the use of structured techniques can reduce the frequency and severity of errors. They explicitly confront the problems associated with unquestioned mental models or mindsets. Investigations therefore require System 2 thinking. Structured analytical techniques are a System 2 approach to overcome System 1 problems.

Fig. 4 Cognitive biases are like optical illusions of thinking: It takes effort to perceive a more accurate reality

2.2.4 Cognitive Biases in Criminal Investigations

The extent of cognitive biases in criminal investigations is well-known, significant, and beyond the scope of this chapter. It should be noted, however, that even the smallest influences can have unintended consequences. For example, in a series of experiments, researchers found that asking "Did you see the broken headlight?" resulted in significantly more—and significantly more confident—reports of having observed a broken headlight instead of asking "Did you see a broken headlight?" The reality was there was no broken headlight and neither group saw one (Loftus and Palmer 1974; Loftus and Zanni 1975). The effects of the broad range of cognitive biases are calamitous and far-reaching (Rossmo 2008). Perhaps the three cognitive biases that are most common are implicit bias, anchoring, and confirmation bias. These three will be briefly addressed in this section.

Implicit bias are unconscious, learned, assessments that act without an individual's awareness or intentional control. The assessments can either positive or negative. Implicit biases lead us to have feelings and attitudes about others based on characteristics like race, ethnicity, age, and appearance. Life experiences, the media, and news are sources of implicit biases and are often reinforced by other biases, like confirmation bias (see below). Implicit biases are universal and pervasive; everybody has them. Oddly, implicit biases may not necessarily align with a person's declared beliefs or even reflect what they say they stand for. However, implicit biases can be changed gradually through a variety of debiasing techniques.

Anchoring is when a person relies too heavily on an initial piece of information offered (the "anchor") when making decisions (Kahneman et al. 2006). The anchor does not even need to be relevant. In one study, people were asked to write the last two digits of their social security number and think about if whether they would pay this amount for merchandise the value of which they did not know. Then they bid for these items in simulated auctions. Those participants with higher two-digit numbers submitted bids that were between 60% and 120% higher than those with lower numbers (Ariely et al. 2003).

Finally, confirmation bias is the search for and favoring of information in a way that supports preexisting beliefs or hypotheses (Mercier 2017). Supporting information is cherry-picked because it reinforces the desired position and any negative information is discarded or disputed. Ambiguous evidence is interpreted as supporting the favored position. Even when the logical support for the belief is completely removed, the bias can persist. For example, one study had participants distinguish between real and fake suicide notes. The feedback on their performance was randomized, with some being told they had done well and others were told they had performed poorly. Afterwards, the participants were told the truth about the experiment and their feedback. Strikingly, participants still thought they were better or worse than average at distinguishing between real and fake suicide notes (depending on the feedback they received) (Ross et al. 1975).

Many biases cannot be eliminated completely but can be mitigated through debiasing techniques, like structured analytic techniques. Documenting the thought process and the evidence supporting it, each hypothesis offered and arguments for and against it, and looking to disconfirm hypotheses instead of piling on cherry-picked information, are the best ways to fight against bias and subjective intuition:

> The best method of avoiding both confirmation bias and misattribution of evidence is for experts to explicitly present the propositions that were considered, and explain how and why the evidence does or does not support each. In this manner, the expert is forced to consider the evidence in light of the various alternatives... while discounting invalid hypotheses. (Aitken et al. 2010)

Sunshine cures many ills. Collaborative interaction produces more hypotheses, a diversity of opinions and viewpoints, and leads to more robust analytic processes. Inaccurate, selective, or unreliable feedback can hamper learning and misleading feedback may increase error rates (Herzog and Fahle 1997). Thus, using methods that provide more accurate feedback directly related to the task (or investigation) at hand should result in improved immediate and future outcomes. Wrongful convictions and bad arrests may be reduced or avoided altogether.

3 Hypotheses and Critical Thinking

3.1 Hypotheses

A fundamental precept of the scientific method is that one should proceed by rejecting or eliminating hypotheses, while tentatively accepting only those hypotheses that cannot be refuted. A hypothesis is a proposed explanation for a phenomenon; it must be a testable statement. The following are examples of hypotheses:

- There is no significant change in my health during the times when I drink green tea only or root beer only.
- Flossing daily prevents cavities.
- All daisies have the same number of petals.
- All swans are white.
- Mr. Johnson killed his wife for the insurance money.
- The jewelry store heist was an inside job.

Hypotheses should be evaluated on a number of qualities, the first being the aforementioned testability. Testability has two aspects. First, the hypothesis must be falsifiable, that is, it has to be possible to disprove the statement through observation or data. Second, it must be feasible to see a series of events to reproduce the proposed explanation. If it is not possible or feasible to test a hypothesis, then it is mere conjecture. Hypotheses can also be judged on their parsimony, how frugal they are with what it would take to be true. In short, simpler is usually better. This quality is summed up in Occam's razor[5]: Among competing hypotheses, the one with the

[5] Named for the English Franciscan friar, William of Ockham (c. 1287–1347), a scholastic philosopher and theologian.

fewest assumptions should be selected. Another quality of a hypothesis is scope: Does it offer a solution to more than one phenomenon? And the last two are fruitfulness, the potential to help explain future phenomena, and conservatism, where the hypothesis "fits" with what is already known. On the last point, caution is appropriate, as biases like confirmation bias can lead one to support a favorite hypothesis that seems to fit the facts over other hypotheses that may have more explanatory power.

The quality of falsifiability is key: It is not possible to universally prove a positive statement. No matter how much information you have that is consistent with a given hypothesis, you cannot prove that hypothesis is true, because the same information may be—and often is—consistent with one or more other hypotheses. On the other hand, all it takes to reject a hypothesis is a single item of evidence that is inconsistent with that hypothesis being true. A classic example is a criminal suspect who has a solid alibi of being somewhere other than the scene at the time of the crime. "All swans are white" can be proven false and is a falsifiable statement, because all it would take is evidence of one black swan to prove it false. Evidence of white swans can be compiled all day, supporting the hypothesis but not conclusively proving it.

People tend to give more weight to information that supports a favorite hypothesis than to information that reduces that support or refutes it; the opposite is the proper course of action. Here is a classroom exercise that helps demonstrate this point. Show a group of people this sequence of numbers:

2,4,6

Ask them to discover the rule that governs the sequence by thinking of three-number sequences that follow what they think is the rule; they should not just blurt out what they think the rule is. Have them ask if their examples conform to the rule. They may offer as many sequences as they like; they should stop offering examples only when they think they have discovered the rule. People will start with other even number sequences, usually by twos or multiples thereof, thinking the rule has to do with increasing the values by 2. The rule is: Any three ascending numbers. This hypothesis can never be proved by enumerating examples of ascending sequences of even numbers that are found to conform to the supposed rule. One can only *disprove* the even-by-two hypothesis by citing an ascending sequence of odd (3, 5, 7) or any set of increasing numbers (2, 181, 1004) and learning that this, too, conforms to the rule; if someone ultimately offers a *descending* example (56, 17, 4), that should cinch the process. People tend to seek evidence that confirms their position, rather than actively try to disprove what they think is true.

The goal here is to refute all plausible alternative hypotheses rather than showing that the favored hypothesis is supported by a lot of evidence. Any hypothesis that cannot be refuted should be taken seriously. The investigator must change perspectives from confirming to refuting, as it forces questions and reveals evidence that would not otherwise be considered. Think of the shift in viewpoints as insurance against unpleasant surprises, like wrongful arrests and convictions.

3.2 *Critical Thinking*

Critical thinking is disciplined thinking that is clear, rational, open-minded, and informed by evidence.[6] The purpose of critical thinking is to make up one's mind about what to believe or do. Engaging in critical thinking attempts to fulfill standards of adequacy and accuracy appropriate to the problem or topic: The thinking fulfills relevant standards to some threshold level. Critical thinking uses evidence through reality, employs skills to isolate the problem from context, uses relevant criteria for making the judgment well, and uses applicable methods or techniques for forming the judgment. The American philosopher and educational reformer, John Dewey (1859–1952) outlined a classic approach to critical thinking with five phases:

1. *Suggestions*, where one or more possible solutions are proposed;
2. Converting the difficulty, challenge, or situation into a *problem* to be solved, forming it into a question so that one or more answers can be considered;
3. Formulating one or more relevant ideas, or *hypotheses*, to frame and guide evidence collection and other materials;
4. *Reasoning*, the mental expansion and elaboration of the collected materials in pursuit of one or more solutions; and
5. *Testing* the hypotheses by actual or intellectual methods. (Dewey 1933, pages 106–107)

The goal of critical thinking is to reach valid conclusions and judge the validity of any inferences. Evaluating multiple, mutually-exclusive hypotheses at the same time provides a more systematic and objective analysis than pursuing one "most-likely" or favorite explanation. Because of their critical role in the criminal justice process and the outsized effects they may have on citizens' lives and liberties, criminal investigations are procedurally and ethically bound to employ critical reasoning. The structured analytical techniques detailed in the next section integrate themselves into this general five-phase process.

4 Structured Analytical Techniques

The techniques offered here are a subset of the ones traditionally taught in the IC (Heuer and Pherson 2015) as not all necessarily apply to criminal investigations; the interested reader would be wise to review the rest for ideas and inspiration. Some changes have been made and some latitude taken with the exact steps or approaches to these methods, largely due to teaching experience. Rigorous adherence to a set of rules often leads to its own type of bias and if a variant of a method works, so much the better; if it does not, ignore it. While some software programs are available that

[6] www.dictionary.com

perform some of the technique leg work, this chapter errs on the side of simplicity, describing the methods using pens, paper, and other normal office supplies.

4.1 *Decomposition and Visualization*

One of greatest constraints an investigator faces is how much information they can keep at the forefront of their minds and think about at the same time. This adds to the investigator's cognitive load, the effort required to concentrate on a task. Two common approaches to coping with this are decomposition, or the breaking down of a problem into component parts, and visualization, arranging all the parts in some organized manner to facilitate understanding how the parts interrelate. Analysis itself is breaking information down into its component parts because anything that has parts also has a structure that relates these parts to each other. As Heuer (2015) notes,

> One of the first steps in doing analysis is to determine an appropriate structure for the analytic problem, so that one can then identify the various parts and begin assembling information on them. Because there are many different kinds of analytic problems, there are also many different ways to structure analysis.

Decomposition methods are used daily, often unrecognized. People used to click the receiver several times to get the operator's attention and then say, "I want to speak with Max on 7th Avenue South". With the invention of the rotary dial in the mid-1920s, people were afraid that misdialed numbers could compromise the system (10% misdials could cause a major failure). A mnemonic system was developed to help people remember numbers. William G. Bauvelt designed the rotary dial system primarily for use of the mnemonic system by grouping the alpha characters in the familiar groups of three per number. The first two letters were used to help remember the exchange, the three number prefix to provide a 7-digit number (PEnnsylvania 6-5000). Area codes and direct full number dialing pushed the mnemonic system out of use. While ten numbers are hard to remember (7278737748), humans can remember about 7 digits, so by chunking the numbers and separating them with a dash ((727) 873-7748), it is easier to recall the string (Miller 1956). If we break the information down to its components and restructure it, we can think about it easier and more clearly.

4.1.1 Linear Ordering

Some of the best techniques are the simplest. Checklists, task lists, to-do lists, troubleshooting lists are all easy, collaborative, and integrate well into the workflow. Hospitals have recently found the benefits of simple checklists: One hospital system employed this approach for catheter lines and in 18 months saved over 1500 lives and $75 million (Gawande 2010). The checklist can be as simple or as process-oriented as desired (Fig. 5).

Chronologies and timelines are both critical and often underused tools in investigative analysis. The difference between them is important. Chronologies arrange events in the order in which they happened: First, this, then that, then the next thing. Recipes are a good example of a chronology ("First add the flour and baking soda. Then mix the sugar and butter until uniform. Next,..."). Timelines are a graphic depiction of those events put in context of the time the events occurred and the time between them; that is, a timeline is events plotted against a quantitative "clock". Timelines are typically linear but scalable; they can be expanded or contracted as required to display events sensibly (Fig. 6). They can be used at the beginning of an investigation to start to make sense of events but also afterwards as a check on hypotheses, with these questions in mind:

- What are the distances between events?
- Were events or processes overlooked?
- Are all the critical and relevant events listed?
- What events outside the timeline could have affected it?
- When did the information become known?

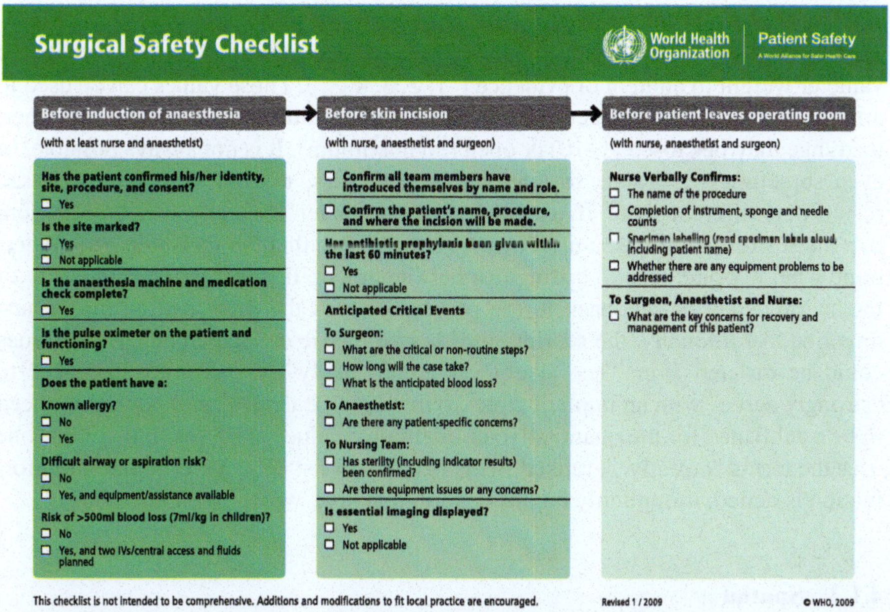

Surgical Safety Checklist

World Health Organization | Patient Safety

Before induction of anaesthesia → **Before skin incision** → **Before patient leaves operating room**

Before induction of anaesthesia
(with at least nurse and anaesthetist)

Has the patient confirmed his/her identity, site, procedure, and consent?
☐ Yes

Is the site marked?
☐ Yes
☐ Not applicable

Is the anaesthesia machine and medication check complete?
☐ Yes

Is the pulse oximeter on the patient and functioning?
☐ Yes

Does the patient have a:

Known allergy?
☐ No
☐ Yes

Difficult airway or aspiration risk?
☐ No
☐ Yes, and equipment/assistance available

Risk of >500ml blood loss (7ml/kg in children)?
☐ No
☐ Yes, and two IVs/central access and fluids planned

Before skin incision
(with nurse, anaesthetist and surgeon)

☐ **Confirm all team members have introduced themselves by name and role.**

☐ **Confirm the patient's name, procedure, and where the incision will be made.**

Has antibiotic prophylaxis been given within the last 60 minutes?
☐ Yes
☐ Not applicable

Anticipated Critical Events

To Surgeon:
☐ What are the critical or non-routine steps?
☐ How long will the case take?
☐ What is the anticipated blood loss?

To Anaesthetist:
☐ Are there any patient-specific concerns?

To Nursing Team:
☐ Has sterility (including indicator results) been confirmed?
☐ Are there equipment issues or any concerns?

Is essential imaging displayed?
☐ Yes
☐ Not applicable

Before patient leaves operating room
(with nurse, anaesthetist and surgeon)

Nurse Verbally Confirms:
☐ The name of the procedure
☐ Completion of instrument, sponge and needle counts
☐ Specimen labelling (read specimen labels aloud, including patient name)
☐ Whether there are any equipment problems to be addressed

To Surgeon, Anaesthetist and Nurse:
☐ What are the key concerns for recovery and management of this patient?

This checklist is not intended to be comprehensive. Additions and modifications to fit local practice are encouraged. Revised 1 / 2009 © WHO, 2009

Fig. 5 The WHO Surgical Safety Checklist was developed after extensive consultation aiming to decrease errors and adverse events, and increase teamwork and communication in surgery. The 19-item checklist has gone on to show significant reduction in both morbidity and mortality and is now used by a majority of surgical providers around the world. https://www.who.int/patientsafety/safesurgery/checklist/en/

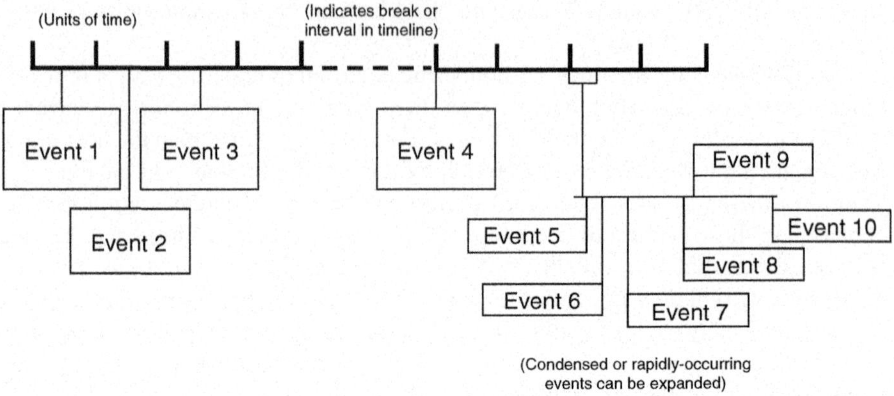

Fig. 6 A timeline is a graphic display of a series of events in chronological order plotted along a quantitative scale with temporal units (minutes, hours, days, weeks, years, etc.)

Sorting is effective when information elements can be broken out into categories or subcategories for comparison. It is particularly effective during the initial data gathering and hypothesis generation phase. The categories are important but changeable; a fluid system of sorting is recommended, such as index cards, sticky notes, white boards, or other similar approaches that allow for additions, changes, and collaboration.

After sorting may come ranking, scoring, and prioritizing. Ranking assigns a value or weight to an item of evidence ("1, 2, 3, 4…"). Those values can be used to arrange evidence into a structure that reflects those values in relation to each other; the range may be closed (1–10) or open (no maximum). It is not always possible, or even sometimes desirable, to have unique rankings, that is, each item assessed results in a different score. If, as will be discussed later, the analysis of two or more hypotheses results in the same score, it is imperative that the next step be to determine what it would take to distinguish between them, if possible. The similarity of the scoring may be feedback to the process or lead the investigation into a new direction. Additionally, the scoring need not be metric or mathematical; the values could be ordered from "less likely" to "more likely" or "strongly disagree" to "strongly agree" with an implicit three- to five-level scale. Rules or definitions need to be established for the quantitative or qualitative values assigned so that everyone uses the terms correctly. A ranked list is just that, a list (Fig. 7). If additional information is added, it implicitly begins to form a matrix, which is addressed next.

4.1.2 Spatial

A matrix is a generic analytic tool for sorting and organizing data in a manner that facilitates comparison and analysis. Matrices are used to analyze the relationships among sets of variables or values. Matrices impose a grid on the data, with as many cells as needed for whatever problem is being analyzed, to arrange it in order and

Rank	Website	Category	Change	Avg. Visit Duration	Pages / Visit
1	google.com	Computers Electronics and Technology > Search Engines	=	00:09:38	8.86
2	youtube.com	Arts and Entertainment > TV Movies and Streaming	=	00:21:48	9.10
3	facebook.com	Computers Electronics and Technology > Social Networks and Online Communities	=	00:11:11	10.77
4	baidu.com	Computers Electronics and Technology > Search Engines	=	00:06:47	7.73
5	instagram.com	Computers Electronics and Technology > Social Networks and Online Communities	=	00:06:28	14.20

Fig. 7 A ranking of the top 5 web pages based on page visits from worldwide sources. Note that the list with the additional information implicitly begins to become a matrix. (Source: www.similarweb.com/top-websites, 7 MAY 2019)

	This item ASUS Chromebook C202SA-YS02 11.6" Ruggedized and Water Resistant Design with 180 Degree (Intel Celeron 4 GB, 16GB eMMC, Dark Blue, Silver) Add to Cart	HP Stream Laptop PC 11.6" Intel N4000 4GB DDR4 SDRAM 32GB eMMC Includes Office 365 Personal for One Year Add to Cart	2018 Acer 15.6" HD WLED Chromebook with 3x Faster WiFi Laptop Computer, Intel Celeron Core N3060 up to 2.48GHz, 4GB RAM, 16GB eMMC, 802.11ac WiFi, Bluetooth 4.2, USB 3.0, HDMI, Chrome OS Add to Cart	ASUS VivoBook W202NA-YS03 Rugged 11.6-inch Windows 10 S K-12 Education Laptop, Intel Dual-Core Celeron processor 2.4GHz, 4GB Ram, 64GB eMMC storage, spill proof keyboard, USB 3.0, HDMI, 2.6lbs Add to Cart
Customer Rating	(1564)	(142)	(147)	(10)
Price	$218.00	$165.88	$204.00	$258.00
Shipping	prime	prime	prime	prime
Sold By	Amazon.com	WorldWide Distributors	Best Electronics (SN Recorded)	Amazon.com
RAM Size	4 GB	4 GB	4 GB	4 GB
Processor (CPU) Manufacturer	Intel	Intel	Intel	Intel
Processor Speed	1.6 GHz	1.1 GHz	1.6 GHz	2.4 GHz
Display Resolution Maximum	1366*768 pixels	1368 x 768	1366 x 768	1366x768 pixels
Screen Size	11.6 in	11.6 in	15.6 in	11.6 in
Display Technology	LED-Lit	—	—	LED-Lit
Hard-Drive Size	16 GB	32 GB	16 GB	64 GB
Item Dimensions	11.6 x 0.9 x 7.9 in	11.8 x 8.1 x 0.7 in	15.1 x 10.1 x 1 in	11.6 x 7.9 x 0.9 in
Item Weight	2.2 lbs	4.41 lbs	4.41 lbs	2.6 lbs
Operating System	Chrome	Windows 10 S	Chrome OS	Windows 10 S
Processor Count	2	1	2	2
RAM Type	DDR3L-1600 SDRAM	DDR4 SDRAM	DDR3 SDRAM	Unknown
Wireless Compatibility	802.11ac	802.11ac, Bluetooth	Bluetooth, 802.11ac	802.11g

Fig. 8 Matrices impose a grid on the data, with as many cells as needed for whatever problem is being analyzed, to arrange it in order and visually. Once the pattern is recognized, matrices appear regularly and often in daily life

visually: Think spreadsheets. Matrices are used routinely in daily life, like retail (Fig. 8). The matrix may be simple, such as a two-by-two sorting matrix (Fig. 9) or more complex, running to hundreds of cells with links, images, or formulae.

Even organized data (text in prose) can be difficult to interpret. For example,

$$t(\text{time}) = 15', T(\text{temperature}) = 32°, t = 0', T = 25°; t = 6', T = 29°; t = 3', T = 27°; t = 12', T = 32°; t = 9'; T = 31°$$

	Urgent	Not Urgent
Important	Unscheduled rework Last minute changes Dealing with late input Forcing decisions & closure	Thoughtful, creative work High-quality reports Training & development Time off
Not Important	Low-value but required reports Non-case emergencies Miscellanious interruptions Administrivia	Over analysis Gossip Idle speculation Web browsing

Fig. 9 Sorting and ranking can naturally lead to matrix analysis. The difference is spatial and the values are spread from a linear list to a spatial component

Fig. 10 Imposing a spatial order on the data reveals the information and its relationships

Time (Min)	Temperature (C)
0	25
3	27
6	29
9	31
12	32
15	32

is indecipherable unless it is placed in a matrix to impose order on the information (Fig. 10). The visualization of the data in a structured fashion reveals the pattern and relationships.

One of the most important matrix analysis methods is a Relationship Matrix. It is closely related to another method discussed later, called Link Analysis, but a Relationship Matrix is an essential step before conducting a Link Analysis. The Relationship Matrix identifies associations between entities, locations, and perhaps objects ("Johnson's vehicle"). Each association should be coded in some way, using symbols, colors, or both (Fig. 11). The entity names, which are the subject of the chart, are entered on the vertical axis of the matrix; people, places, and things may be grouped. Individuals and organizations should be listed in alphabetical order, with the individuals listed first. Alternately, the individuals could be coded, say, by color, by organization. Vehicle Identification Numbers (VINs), addresses, and other locations should be arranged in alphanumeric order after the organizations. If non-numeric values are used, a simply tallying of the number of associations can be helpful to demonstrated the relatedness of the person, place, or thing. The goal is to make the relationships apparent and to simplify the data, so a complicated or visually dense coding scheme will not be useful (Fig. 12).

After the Relationship Matrix, a further visualization is a Link Analysis, which diagrams the relationships between a number of people and organizations in a visual

Fig. 11 Association codes are used to indicate the basis for or nature of each relationship shown in the matrix. The symbols or values must make the relationships clearly apparent

Association Codes

Code	*Meaning*
●	Confirmed, personal
○	Suspected, personal
+	Confirmed, organizational
--	Suspected, organizational
▲	Confirmed, business
△	Suspected, business

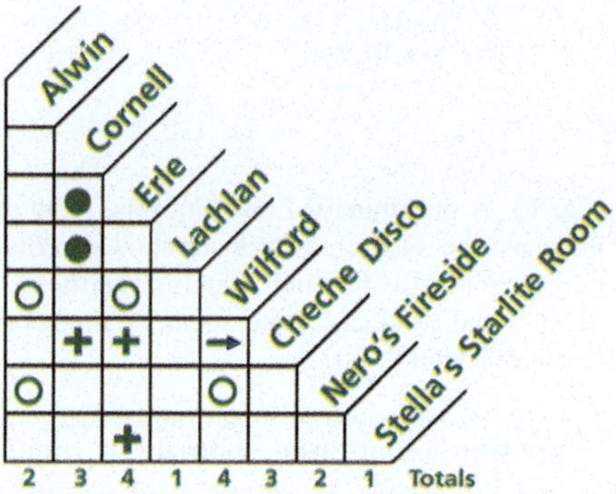

Fig. 12 A completed Relationship Matrix. (Source: UNODC 2011)

form beyond a matrix layout. Analyzing links has a history in criminal investigations as "murder boards," a corkboard with photographs, notes, and other items physically pinned to it with strings running between the related items. An organizational chart is a more common example of a Link Analysis. The investigator studies reports and identifies the people, objects, and organizations pertinent to the investigation. Typically, confirmed links are shown in solid lines while suspected ones are shown in dotted lines. Individuals within an organization may be drawn together inside that designation (suggesting a conceptual method, the Venn Diagram, discussed in the next section) with links drawn to both the individuals within that organization and the organization itself (Fig. 13).

All charts should be timed, dated and numbered as to sequence of occurrence. A Relationship Matrix and a Link Analysis may only show a snapshot of the situation and it may be more.

Meaningful to show the evolution of the relationships over time. Each chart or the series should have a key to explain symbols and other indicators.

The UNODC (2011) notes that the important features of a network are:

- **Entities,** the items under investigation which could include people, businesses, organizations, means of transport, locations, events, objects etc.
- **Relationships** can be familial or based on give and take such as reciprocity (exchange and compromise), suitability (the right person to do the job), bonding

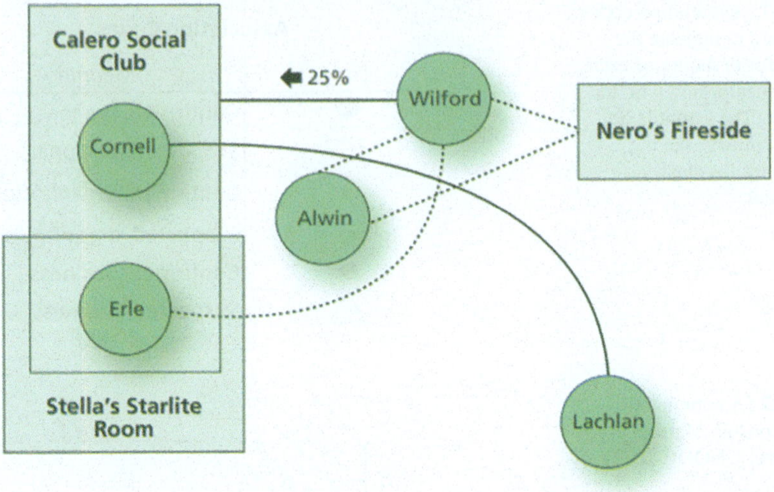

Fig. 13 A preliminary Link Analysis. Note that the confirmed relationships are shown in solid lines and the suspected ones are shown in dotted lines. Also, individuals within an organization may be linked to other individuals or entities as well as the organization to which they belong. The "25%" could indicate time spent or financial investment but should be explained, regardless. (Source: UNODC 2011)

(past associations), control (criminal hierarchy or threat), predominance, superiority, subordination and succession.

- **Directionality** relates to the flow of information, favours and authority and enables understanding of the internal mechanics of a network.
- **Strength** is a subjective judgement based on interactions included in the relationships and evaluation of the data provided.

A Link Analysis is a work product from which hypotheses could be generated regarding the status of associations among people, places, things, like organizations. The analysis should include evaluating relationships, their links, their strengths and directions, what those links mean to those investigating a case or cases. The links may be updated as new information becomes available.

4.1.3 Conceptual

Venn diagrams show all possible relations between a selection of different sets represented by multiple overlapping circles (Fig. 14). Figure 14 shows three sets, A, B, and A&B. Not all elements of set A intersect with elements of set B and vice versa; only those elements that do intersect are contained in the set A&B. Venn diagrams are useful to show groups, affiliations, or to arrange relationships. For example, of all the items stolen from a residence (A), only some are found (A&B) in a particular pawn shop (B). Venn diagrams are easy to draw and are readily updated.

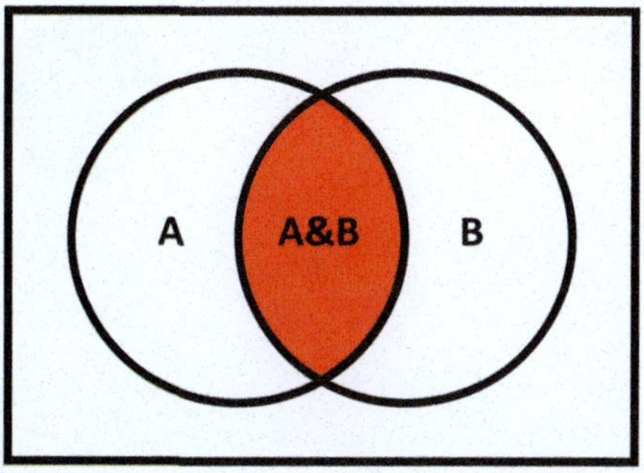

Fig. 14 Venn diagrams show the logical relations between sets. They were invented by John Venn around 1880

Another type of concept visualization is generically known as a mind map. Mind maps are visual representations of how an individual or a group thinks about a topic of interest. They are composed of the ideas and the links between those ideas. They help sort out thinking, facilitate the communication of a complex set of relationships to others, and to identify gaps or stimulate new thinking on a topic. Mind maps can be constructed using whiteboards, index cards, sticky notes, and free or paid software (Fig. 15).

4.1.4 Activity

Process maps are used in business, quality assurance, and the military to diagram each step in a complicated or complex process. Process maps are useful in criminal investigations, as well, but rather than using the tool to map out steps to achieve a future goal, they are useful in diagramming an event that has already happened. The map may be plotted as a chronology (steps in order) or as a timeline (steps against the scale of time) (Fig. 16). Maps may be created manually or with software. The events may contain brief (three or four words) descriptions of the events. Consistency in using colors or symbols is important (green for financial steps, for example). Lines connect the related events and arrows show direction. Formatting of process maps is very flexible and the components can be combined in a variety of ways. The main points for making a process map are to provide a clear and accurate presentation of the information and to keep the chart as simple and to the point as possible (UNODC 2011).

4.2 Structured Brainstorming

True creativity is a reactive event, where a person responds to stimuli or events that they encounter. In criminal investigations, creativity typically comes from two or more people bouncing ideas off each other. These can be discussions or, more formally, "brainstorming" sessions that expose one to a greater range of ideas and

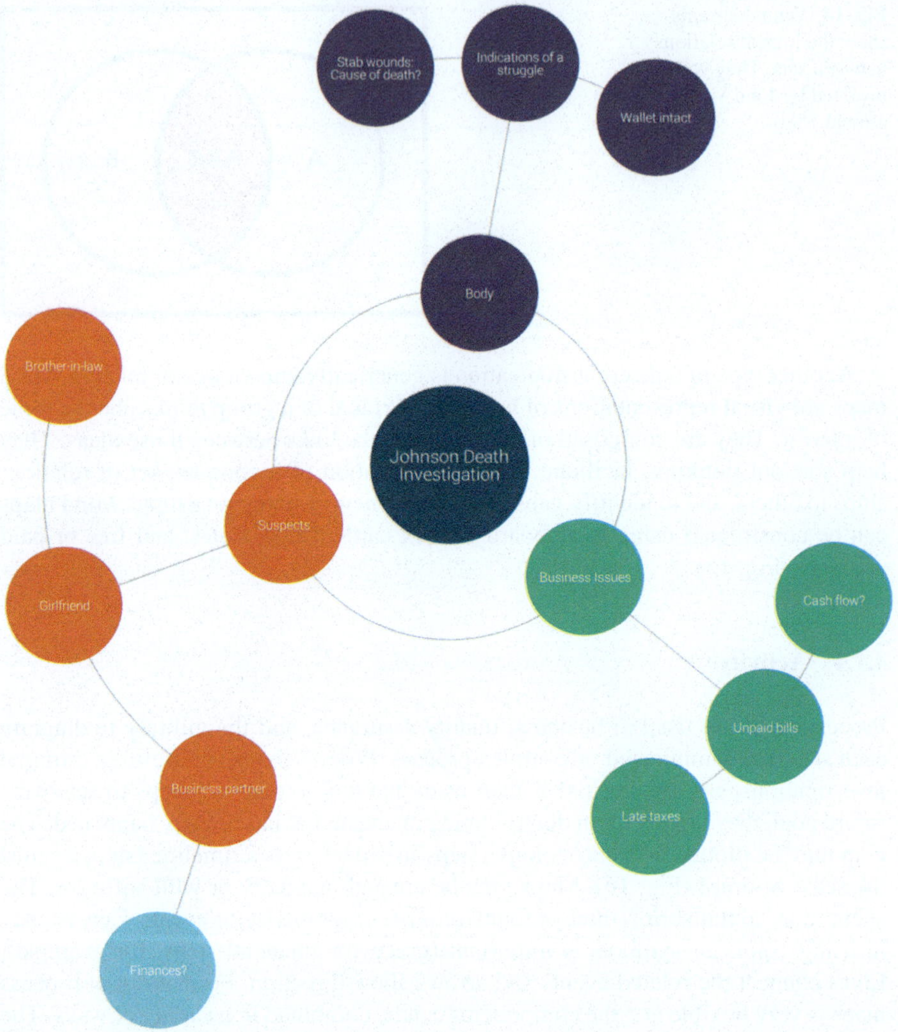

Fig. 15 A mindmap is a visual display of conceptual relationships

perspectives than doing it alone. The goal is to achieve a broadened perspective which in turn produces a better analytical product. Generally, the rules for brainstorming are simple enough and seem egalitarian: People personally offer ideas vocally without criticism. The problem with this approach is that traditional brainstorming doesn't work. Studies show that participants feel pressured to go along with the dominant pattern of thinking ("collaborative fixation"), leading to conformity of ideas and reduction of creativity. Early ideas tend to have disproportionate

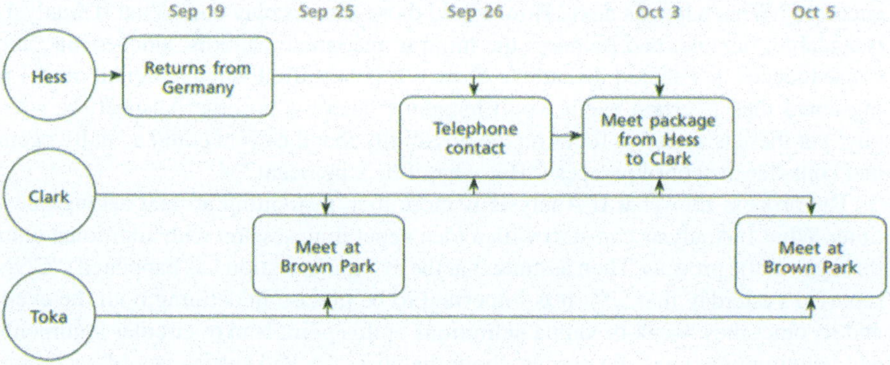

Fig. 16 A timeline process map. (Source: UNODC 2011)

influence over the rest of the conversation ("anchoring"). In most meetings with traditional brainstorming, a few people do 60–75% of the talking.[7]

In reality, idea generation should be separate from discussion to avoid biasing effects, like anchoring or the bandwagon effect. Allowing constructive criticism and light dissent can lead to greater ingenuity. One approach, "brainwriting," provides for a more objective and safe creative environment. The goal of the session is written at the top of a board by a session leader, along with a list of resources to achieve that goal. The participants then write down as many ideas for tackling the problem as possible; this allows for both "top-down" and "bottom-up" thinkers to go their own ways. Writing questions rather than solutions is also a productive approach ("What is keeping us from…", "Why can't this be…"). Individuals can be paired up to revise and improve questions collaboratively. The ideas are then shared in a structured way, going through each idea. At the end, votes are cast for the best ideas (ranked by number or weight of votes using a show of hands, stickers, or other indicators). No guessing or confessing to authorship of ideas is allowed.

4.3 "What If" Analysis

What if Analysis is a technique that can be used throughout an investigation, not only at the beginning. Interestingly, it can also be extremely useful at the end (or the perceived end) of an investigation. What If Analysis imagines that a significant event has occurred and with the benefit of the artificial "hindsight," the analyst conjectures how it might have happened. For example, imagine Suspect A has been arrested but none of the forensic test results that come back after the arrest support Suspect A's guilt. What would be necessary and sufficient for the event to have

[7] Brooks, L. n.d. "Brainstorm Questions, Not Solutions", at: http://99u.com/articles/52335/brainstorm-questions-not-solutions

occurred? What happens next? How would those actions play out? What If Analysis is scalable, flexible, and prepares the mind to recognize patterns, antecedents, and consequences in a shared discursive format. It is best used when a mental model is ingrained that a certain event won't happen ("There is no other suspect; he's *the* guy") or there is a need to focus on the possibility that the event could actually occur and consider the consequences ("But what if he's innocent?").

The process for What If Analysis is more than just sitting around asking questions. What If Analysis can start with a idea generating session with additional ones throughout the process. Then assume that the event in question has happened ("CNN reported yesterday that…"). It is important to be precise in defining both the event and its outcomes; weak or vague definitions at this point lead to circular arguments and hedging outcomes. Be certain about the necessity and sufficiency of each item (oxygen, heat, and fuel are all necessary for fire but only all three are sufficient to create it). Keep the variables minimal and conservative; complexity is sometimes a necessity with any analysis, but it can also be an enemy and a What If Analysis can get out of hand quickly. The what-if analysis is about asking questions. But too many data alterations to satisfy these questions along the way end up making the outcome tougher to explain. Also, if possible, keep the variables in the near term. The further out in time variables are proposed, the less likely the environment and circumstances around that question will be accurate and aligned with the current situation.

Define any follow-on effects and further outcomes; challenge assumptions (see Key Assumption Check) as you go. As the process develops, a chain of argumentation needs to develop to explain how these outcomes could have come about. Assess and rate the severity of each outcome and what would it take to avoid, mitigate, or prevent it; rank order the scenarios according to how much attention they merit to better allocate resources. Finally, work backwards against the current status looking for evidence gaps or logic faults—where and how could the current status be wrong?

What If Analysis is a logical follow-up to the Key Assumptions Check and Analysis of Competing Hypotheses because it shifts the focus from "how likely is it" to:

- How could it be possible?
- What would be *necessary* for this to happen?
- What would be *sufficient* for this to happen?
- What would be the consequences?
- Has the possibility of this happening increased?

What If Analysis is a good preventative tool if used as feedback because it provides specific indicators to monitor for change (If another similar crime is committed, is the correct suspect in jail? Was there an accomplice?).

4.4 *Morphological Analysis*

Imagine being in a dark room. The room's walls or borders need to be found to understand the size and shape of the room, as well as any obstacles (furniture, lamps, etc.). Morphological Analysis is a technique for defining the size, shape, and complexity of complex, non-quantifiable problems for which little information is available and the chances of surprise are great. Morphological Analysis is best used early in an investigation to identify a range of possibilities without getting bound up in all the details of each one. By generating a comprehensive list of possibilities, investigators are in a better position to identify and select the most credible outcomes. Investigators can then take necessary actions to follow the best leads to greatest effect, either advancing the investigation or reducing waste of time and resources. Morphological Analysis leaves investigators with not only a clear set of potential plans and necessary leads to follow, but also a clear audit trail about what was decided and why in case of inquiries or oversight.

Morphological Analysis works as follows. A key set of parameters or dimensions of the problem are defined. Each of those parameters are broken down into relevant forms, states, or values. Beyond two dimensions (like a matrix), multiple dimensions may be needed (software or more pages of paper). Often, it is possible to discard the majority of possibilities as not physically possible, impracticable, or dismissable; they should not be deleted as they may become viable at a later date.

An example of Morphological Analysis might be a taking a vacation. Where should the vacation be taken: The beach, a major city, the mountains? Each of those is a state or value of the parameter "location." Should you go alone, with a significant other, or with family? What season will you go? Figure 17 demonstrates an example of a Morphological Analysis for a vacation.

The various combinations of values can then be explored and a What If Analysis could be conducted ("A summer vacation in New York city for just the two of us to site-see could be reasonably priced; what if we did it in the fall?"). Similarly, in a criminal investigation, like a carjacking, a range of possibilities can be developed and each set of values explored (Fig. 18).

4.4.1 Indicators

One outcome of a Morphological Analysis may be one or more indicators, specific, observable and measurable characteristic that can help track events, spot emerging trends, and warn of unanticipated changes. Indicators are pre-established set of actions, conditions, facts, or events whose co-occurrence argue strongly that a phenomenon is present or likely to occur. The indicator should be focused, clear and specific; an indicator should be defined in precise, unambiguous terms that describe clearly and exactly what is being measured. There should be at least one indicator for each outcome.

- Where
  - Beach
  - Mountains
  - City
- Purpose
  - Relax
  - Adventure
  - Site-see
- Who's going
  - Just me
  - Me and significant other
  - Group
- When
  - Summer
  - Winter
  - Fall
  - Spring
- Cost
  - Cheap
  - Medium
  - Luxury
- Geography
  - Domestic
  - International

Fig. 17 A set of key parameters are defined (Where, Purpose, Who's Going, When…) and for each one, a range of values or states are listed (Beach, Mountains, City). This way, the various possibilities can be listed in a way that groups them into relevant categories but still provides all combinations

- Where
  - Parking lot/garage
  - Gas station
  - Traffic intersection
- Value of car
  - Low
  - Medium
  - High
- Number of perpetrators
  - One
  - More than one
- Weapon
  - Yes
  - No
- Injury
  - Driver
  - Perpetrator(s)
  - Occupant
- Occupants still in car
  - Yes
  - No

Where	**Parking lot**	Gas station	Intersection
Value of car	Low	**Medium**	High
# Perp	**1**	2+	
Weapon	Yes	**No**	
Injury	Driver	Occupant	**Perpetrator**
Occupant	Yes	**No**	

One scenario is where a single perpetrator without a weapon carjacks a medium value vehicle in a parking lot; only the driver is in the car and the perpetrator is injured during the crime.

Where	**Parking lot**	Gas station	Intersection
Value of car	Low	Medium	**High**
# Perp	1	**2+**	
Weapon	**Yes**	No	
Injury	**Driver**	Occupant	**Perpetrator**
Occupant	Yes	**No**	

Another scenario is where multiple perpetrators carjack a high-value vehicle using a weapon where both the carjacker and the driver are injured; no one else is in the car.

Fig. 18 Once the range of possibilities is developed, each value-set can be explored, assessing approaches, ideas, plans, resources, and outcomes

Humans sometimes see what they expect to see: Indicators create an awareness that prepares the mind to recognize warning signs or flags beyond expectations. Change often happens so gradually that investigators don't see it or they rationalize it as not being of importance until it is too late; by specifying in advance the criteria or threshold for what is significant, investigators can avoid certain cognitive biases, including anchoring and entrenchment. Good indicators are observable and collectible, valid, reliable, timely, and specific to the scenario of interest.

As enhancements, a rating or ranking (see above) for each indicator can be established to weight their evaluation or potential. A schedule can be created to review and rate the indicators periodically. A narrative description for each point on the scale, describing what would be expected at that level, should be established. Any sources used to generate the rating should be listed and verified.

4.5 Key Assumption Check

People make and use assumptions based on learned experiences, whether what they learned was right or wrong. More education and more experience are often thought to be beneficial (and they can be) but danger lurks in too much unvalidated experience. As individuals progress through learning, it is commonly said they pass through four phases[8]:

1. Unconscious incompetence: They do not know what they do not know.
2. Conscious incompetence: They know what they do not know.
3. Conscious competence: They know what they know.
4. Unconscious competence: They do not know what they know.

People are most dangerous at the first and last stages. Those at the low end of competence are generally unaware of how incompetent they are (Kruger and Dunning 1999). At the unconscious competence stage, unvalidated experience can lead to assumptions that are unjustified and perhaps wrong; the same mistake can be made for years and it is still considered "experience." Moreover, each person's experience is different and usually unstated, leading to an untenable, inaccurate situation:

> Accuracy of interpretation never can be supplanted by assumption. I assume one thing, you assume something else, and the other fellows each assume something different from each of the others. Grant the acceptance of assumption, and then any one of a 100 different solutions is correct. Could anything be more ridiculous? (Bennet 1933, pg 158)

Therefore, the specific assumptions that underpin the basic analytic line of reasoning need to be made explicit, evaluated, and discussed for any investigation to grounded in reality (Nkwake 2013). The technique called Key Assumption Check (KAC) is designed to do that and more, improving the foundation of the inquiry. The technique adds the following value to an investigation (Heuer and Pherson 2015):

[8] The model was developed at Gordon Training International in the 1970s with different terminology than is commonly used now (gordontraining.com). The model is frequently but incorrectly attributed to Abraham Maslow. It is conceptually similar to the Johari Window (Luft and Ingham 1961) which lists knowns and unknowns for an individual in relation to others in a heuristic exercise.

- Improves the understanding of the current baseline interactions
- Clarifies perspective and encourages thinking about the issues
- Reveals hidden relationships that might have been ignored or overlooked
- Mitigates surprises in the face of new disconfirming information

Similar to brainstorming, a list of assumptions is brought by each person to a meeting. The assumptions are presented for all to see and read (anonymity may be useful). Any additional assumptions are added, without critiquing the new or current ones. The universal slate of questions is asked regarding each assumption (Who, what, where, when, how, why). It is important to remember that many key assumptions will turn out to be key uncertainties or key biases. Any absolute or causal statements ("will always," "never," "would have to be," among others) should be challenged. At the end of the session, a key question to ask is "What else is so obvious that we're not challenging it?," which may be difficult to address because of the assumed obviousness of the statement.

Each assumption should be characterized in some way, for example:

- Confirmed
- Correct with caveats
- Questionable

Or, using the acronym FOG,

- Fact: Referenced, confirmed
- Opinion: Supported, not conclusive
- Guess: Estimate (or SWAG[9])

By characterizing each assumption and grounding it more or less in validity, the assumptions can be ranked and valued for later use and possible reevaluation. The list of assumptions is then refined and assumptions that do not hold up to scrutiny are removed from the active list. Any assumption that would, if wrong, lead to changing one or more conclusions should be highlighted for future consideration. The strength of an analytic chain of reasoning is no greater than its weakest link. An example of a Key Assumption Check is shown in Fig. 19.

Although intended to be used in an early phase of an investigation, KACs should be made available to every team member and revisited occasionally as new information comes to light or the situation changes.

4.6 Analysis of Competing Hypotheses

Analysis of Competing Hypotheses (ACH) is a visual assessment of testable statements given certain items of evidence. ACH is intended to assist investigators in making judgments on issues that require weighing alternative explanations; there-

[9]A slang military phrase for Scientific Wild-Ass Guess (Safire 2004).

Key Assumptions Check: The 2002 DC Sniper Case

The outbreak of sniper shootings in the Washington, DC area in the fall of 2002 provides a good example of how this technique could have been applied. After the initial flurry of shootings, the operating assumption that quickly emerged was that the shootings were the work of a single, white male who had some military training and was driving a white van. If law enforcement officials had conducted a Key Assumptions Check, they could have broken this statement into its key components and assessed the validity of each statement as follows:

Key Assumption	Assessment
The sniper is a male.	Highly likely (but not certain) given past precedent with serial killers. We are taking little risk by not looking for a female.
The sniper is acting alone.	Highly likely (but not certain) given past precedents.
The sniper is white.	Likely, but not as certain, given past precedents. We would be taking some risk if we rule out nonwhites as suspects.
The sniper has military training/experience.	Possible, but not sufficient reason to exclude from consideration potential suspects who have not had any military training.
The sniper is driving a white van.	Possible because you have a credible eyewitness account but worthy of continuing scrutiny given the number of white vans in the area (more than 70,000 registered in the Maryland suburbs of Metropolitan Washington, DC) and that different kinds of vehicles are being described.

A Key Assumptions Check could have allowed law enforcement officials to:

- Avoid jumping to conclusions (the sniper is white, has military training, and is driving a white van) that did not hold up under closer scrutiny. By explicitly examining each assumption, officials could have avoided prematurely narrowing down the potential pool of suspects to a group that did not include the actual perpetrator. Similarly, they might have been more cautious about accepting that the sniper was driving a white van.
- Be receptive to new leads and citizen tips, such as eyewitness reports that the sniper fled the scene driving a specific model Chevrolet.
- More seriously consider evidence that subsequently became available, which contradicted a key assumption. If officials had stated explicitly that they were assuming the sniper was acting alone, they might have been sensitive to new information that contradicted that key assumption. Often this type of information gets "lost in the noise" if the analyst has not already thought about what key assumptions he or she is making.

Fig. 19 A Key Assumption Check product on the DC Sniper Case. (Source: CIA 2009)

fore, ACH is applicable in many disciplines, not just intelligence or criminal investigations. *The core idea behind ACH is to refute rather than confirm each of the hypotheses*. Any statement can be confirmed by cherry-picking evidence or twisting facts to support it. As discussed earlier, falsifiability of a statement is central to its strength as a hypothesis and actively seeking evidence that would prove a statement wrong is the most objective approach to take. Thus, the most likely hypotheses are the ones with the *least* evidence *against* them. Note the use of the plural: ACH requires the generation and testing of multiple hypotheses rather than the support of a single one. ACH prompts investigators to develop a full set of alternative hypotheses, producing a more comprehensive range of potential explanations and reducing the risk of satisficing, accepting the first answer that seems to fit rather than an optimal one (Simon 1956). ACH acts as a guard against confirmation bias. The approach, whether by hand on paper or through the use of software, helps to maintain a record of relevant evidence and how that evidence relates to each hypothesis. Finally, ACH organizes conclusions and makes reasoning transparent, providing an audit trail, demonstrating how conclusions were reached, whether the auditor agrees with them or not. In that sense, an ACH is a kind of decision matrix (Pugh 1991).

The outcome of an ACH is only as good as the hypotheses proposed and the evidence collected (garbage in, garbage out). Evaluating evidence requires an assessment of the validity or accuracy of the source of that evidence.

4.6.1 Assessing the Validity of Evidence

The system for assessing the source and intelligence reliability was developed around the time of World War II. The reliability of the source is evaluated separately from that of the information. The source reliability is rated between A (history of complete reliability) to E (history of invalid information), with F for source without sufficient history to establish reliability. The information content is rated between 1 (confirmed) to 5 (improbable), with 6 for information whose reliability can not be evaluated (see Fig. 20).

SOURCE Reliability

Rating		Description
A	Reliable	No doubt about the source's authenticity, trustworthiness, or competency. History of complete reliability.
B	Usually reliable	Minor doubts. History of mostly valid information.
C	Fairly reliable	Doubts. Provided valid information in the past.
D	Not usually reliable	Significant doubts. Provided valid information in the past.
E	Unreliable	Lacks authenticity, trustworthiness, and competency. History of invalid information.
F	Reliability unknown	Insufficient information to evaluate reliability. May or may not be reliable.

INFORMATION Reliability

Rating		Description
1	Confirmed by independent Sources	Logical, consistent with other relevant information, confirmed by independent sources.
2	Probably true	Logical, consistent with other relevant information, not confirmed.
3	Possibly true	Reasonably logical, agrees with some relevant information, not confirmed.
4	Doubtful	Not logical but possible, no other information on the subject, not confirmed.
5	Improbable	Not logical, contradicted by other relevant information.
6	Difficult to say	The validity of the information can not be determined.

Fig. 20 The source reliability is rated between A (history of complete reliability) to E (history of invalid information), with F for source without sufficient history to establish reliability. The information content is rated between 1 (confirmed) to 5 (improbable), with 6 for information whose reliability can not be evaluated. (USDOD 2017)

A reliable source which provides confirmed information would be rated A1, a new source with information that can not be confirmed would be a F6, and a fairly reliable source with logically consistent but unconfirmed information would be rated C2. While the system was intended for human sources, it can be applied broadly to other types of sources and information.

4.6.2 The ACH Process

Through whichever media are chosen to conduct an ACH, the goal is to step through a structured thought process. As in science, reproducibility is an important aspect of a valid reasoning and outcome. To demonstrate the ACH process, the fictitious example of the AAA Junkyard Investigation will be used. In an investigation into stolen vehicle parts, a business, the AAA Junkyard, has come under suspicion. The following facts are known:

- AAA Junkyard is owned by Bill and Sam.
- Vehicle parts have been sold to an undercover investigator (UCI) as part of a sting operation; the investigator knew they were stolen by the serial numbers on the parts.
- Surveillance has observed Bob (non-employee) bringing in what appeared to be stolen parts to AAA Junkyard.
- Joe (non-employee) was also in the AAA Junkyard when Bob delivered the alleged stolen parts.

The steps listed are derived from Heuer (1999) and Heuer and Pherson (2015).

4.6.2.1 Define the Question to Be Addressed

All individuals involved in the process must agree on the question to be answered. It is best to have it written down and have each person sign or initial the statement to document agreement of the question to be dealt with. What is the question to be answered in the example? The most likely response is whether or not the AAA Junkyard is dealing in stolen vehicle parts (that is, is it a "chop shop?").

4.6.2.2 Identify All the Possible Hypotheses, Making Sure They Are Mutually Exclusive

Through brainstorming, Morphological Analysis, and What If Analysis, a list of comprehensive and mutually exclusive hypotheses is developed. Mutual exclusivity is important because, logically, if any of the hypotheses is true, then all of the other ones must be false. While this may seem like a restriction, this rule can lead to new hypotheses: If Hypothesis 1 states that Linda committed the crime and Hypothesis 2 states the Robert committed the crime, they are not mutually exclusive as Linda and Robert could have collaborated on the crime, leading to Hypothesis 3. For the AAA Junkyard Investigation, here are some hypotheses (H):

- H1: AAA is a chop shop (takes cars apart to sell)
- H2: AAA is not a chop shop (honest)
- H3: AAA only sells stolen parts (pass-through)
- H4: AAA was caught in a bad sting operation

Another approach could be to take it to the suspect level:

- Bill is unaware of stolen parts being sold at AAA Junkyard
- Bill is aware of stolen parts being sold at AAA Junkyard
- Sam is unaware of stolen parts being sold at AAA Junkyard
- Sam is aware of stolen parts being sold at AAA Junkyard
- Bob is unaware of stolen parts being sold at AAA Junkyard
- Bob is aware of stolen parts being sold at AAA Junkyard
- Joe is unaware of stolen parts being sold at AAA Junkyard
- Joe is aware of stolen parts being sold at AAA Junkyard

While these might be seen as not mutually exclusive, the hypotheses do present statements for each suspect which are either true or false; some combination of results could lead to an improved investigative strategy ("What does Joe know?").

4.6.2.3 Make a List of Significant Evidence and Arguments for and Against All the Hypotheses

List the evidence that is known; remember to evaluate each item or statement. Assumptions, facts, conclusions from other analyses (like forensic results) should be included. Listing what evidence would be expected to be absent if a hypothesis were true is also important.

4.6.2.4 Prepare a Matrix to Analyze the Diagnosticity of the Evidence and Arguments

Create a matrix (table) with the hypotheses across the top and the evidence listed along the side; typically, there will be more evidence than hypotheses. Code the validity of or evaluate each item of evidence in the context of each hypothesis (Fig. 21). Using Consistent/Inconsistent/Refutes, FOG (Fact, Opinion, Guess), numerical values (either positive or signed, like −3 to +3); an alternative would be to rate the consistency and then add a weight to it based on whether it is fact, opinion, or guess. Color coding, like using a traffic light scheme (red, yellow, green), is visually appealing, as well. An example ACH for the AAA Junkyard Investigation is shown in Fig. 22.

Evidence	Key Assumptions	Competing hypotheses					
		H1	H2	H3	H4	H5	H6
E1							
E2							
E3							
E4							
E5							
# of inconsistencies							

Fig. 21 An ACH matrix, listing hypotheses across the top and evidence along the side. Values are entered into cells and then tallied at the bottom

Evidence	FOG	Key Assumptions	Competing Hypotheses					
			H1	H2	H3	H4	H5	H6
E1 Joe was present during Bob's delivery	F	Joe is involved in AAA Junkyard's business				4	3	
E2 Bob delivered what appear to be stolen goods	O	Bob may be supplying AAA Junkyard with stolen goods to sell	5	5	2	4		5
E3 Bill & Sam sold stolen parts to UCIs	F	Bill & Sam knew the parts were stolen	5	5	2			4
E4 Bill & Sam own AAA Junkyard	F	Bill & Sam control and are involved in their business's dealings	4	4	3		3	4
E5 AAA Junkyard was the focus of sting operation	F	Something about AAA Junkyard caught the attention of law enforcement	5	5	3	3	2	5
		Total Consistencies:	19	19	10	11	8	18

Weight	
5	Very Consistent
4	Slightly Consistent
	N/A
3	Slightly Inconsistent
2	Inconsistent
1	Refuted

Hypotheses
H1: Bill & Sam own/operate a chop shop.
H2: Bob knowingly brought in stolen car parts.
H3: Bob did not know the parts he delivered were stolen.
H4: Joe and Bob know each other.
H5: Joe is not involved in any business.
H6: AAA Junkyard is part of a larger criminal organization.

Fig. 22 An example of an ACH on the AAA Junkyard Investigation, using quantitative and color-coding to help evaluate the hypotheses. (Source: H. Sherman, with permission)

4.6.2.5 Refine the Matrix and Reconsider the Hypotheses

While it may seem static, an ACH is a dynamic, updatable technique to assist with investigations. Has the process so far changed your view on any of the evidence or hypotheses? Are the evaluations (qualitative or quantitative) consistent or are there discrepancies between you and your colleagues? Is there any evidence that seems to be less diagnostic than before? If so, move it to the bottom of the matrix or shift it to another list (but don't delete it). Do any gaps exist between items of evidence and the hypotheses?

4.6.2.6 Compare your Personal Conclusions About the Relative Likelihood of each Hypothesis with the Inconsistency Scores

The most likely hypothesis is the one that is least refuted. Re-sort the hypotheses from least to most refuted. Re-order the evidence (separate list) to see what is most diagnostic. Now would be a good time to conduct a What If Analysis on your least refuted hypothesis. What would it take for it to be wrong? What evidence would refute it? Actively seek that evidence out and, if it exists, find it. Are any of the conclusions ambiguous? Are more than one hypothesis "least refuted" (that is, do they have the same scores)? If so, refine them by deciding what evidence would be necessary to distinguish between them and then seek it out and add it to the ACH.

4.6.2.7 Report your Conclusions

Prepare a draft briefing of your conclusions, explaining the relative likelihood of all the hypotheses, not just the least refuted one. Formats vary by agency but generally a brief should contain the following sections (Major 2008):

- **Bottom Line Up Front (BLUF)**: One or two sentences that summarize the core elements of the events, evidence, and outcomes
- **Evidence/Events**: Who, what, when, where, why, and how (as appropriate)
  - What events happened
  - Dates and times
  - What facts or information provide evidence or support
- **Narrative/Discussion/Interpretation**: The story of what happened in prose form. While the BLUF may change depending on the audience, the narrative won't materially change.
- **Outcome**: What was the result(s) of what happened? This section may include policy implications, depending on the audience.
- **Sources**: References, web sites, news articles, etc.

Include visuals of any analytic technique products, like matrices, charts, timelines, etc.

4.6.2.8 Identify Indicators or Milestones for Future Observation

See Sect. 4.4.1 about creating indicators for milestones (showing progress) or flagging potential issues or leads that suggest a hypothesis might be changing or that new evidence has been found. Remember, the ACH is dynamic and should be updated and revised as new information becomes available.

5 Integrating Structured Analytical Techniques into Criminal Investigations

The intelligence community has used structured analytic techniques for years. The techniques are designed to push analysts out of habitual thinking and away from biases in order to increase creativity, better evaluate questions, and create a document trail that reveals the thinking process that led to the final product. Adapting these methods for criminal investigations should help reduce bias and improve accuracy. Beyond the implications for improving criminal investigations from an agency standpoint, the larger moral and financial value of better investigations justifies this approach. Known wrongful convictions increase almost daily: The Innocence Registry counts almost 2500 exonerations to date, making for nearly 22,000 years lost to family, friends, and society. Jurisdictions pay out reparations to exonerees (more than $2.2 billion to date) and more are to be paid. If that $2.2 billion, minus the operational costs to conduct accurate investigations, had been invested in policing, community programs, or other preventative measures, imagine how much crime would have been prevented and how much goodwill with the public would have been gained.

Structured analytical techniques are not a panacea by any means. But, even if they were no more accurate than current intuitive methods, the transparency and document trail they leave would make case reviews and cold case investigations easier, more accurate, and less fault-finding. They are simple to use, inexpensive, and largely visual. By promoting transparency, creativity, and group discussion, structured analytic techniques can lead to better-supported results in criminal investigations and improved justice for all.

References[10]

Aitken, C., Roberts, P., & Jackson, G. (2010). *Fundamentals of probability and statistical evidence in criminal proceedings: Guidance for judges, lawyers, forensic scientists and expert witnesses*. London: Royal Statistical Society.

Ariely, D., Loewenstein, G., & Prelec, D. (2003). "Coherent arbitrariness": Stable demand curves without stable preferences. *The Quarterly Journal of Economics, 118*(1), 73–106.

Artner, S. J., Girven, R., & Bruce, J. B. (2016). *Assessing the value of structured analytic techniques in the US intelligence community*. Santa Monica: Rand Corporation.

Bennett, G. E. (1933). Assumptions. *The Accounting Review, 8*(2), 157–159.

Burns, B., & Corpus, B. (2004). Randomness and inductions from streaks: "Gambler's fallacy" versus "hot hand". *Psychonomic Bulletin & Review, 11*(1), 179.

Central Intelligence Agency. (2009). *A tradecraft primer: Structured analytic techniques for improving intelligence analysis*. Central Intelligence Agency: Langley.

Dewey, J. (1933). *How we think*. Buffalo: Prometheus Books. (Original work published 1910).

Dror, I. E., & Hampikian, G. (2011). Subjectivity and bias in forensic DNA mixture interpretation. *Science & Justice, 51*(4), 204–208.

Dror, I. E., Charlton, D., & Péron, A. E. (2006). Contextual information renders experts vulnerable to making erroneous identifications. *Forensic Science International, 156*(1), 74–78.

Einhorn, H. J., & Hogarth, R. M. (1978). Confidence in judgment: Persistence of the illusion of validity. *Psychological Review, 85*(5), 395.

Ericsson, K. A., Hoffman, R. R., Kozbelt, A., & Williams, A. M. (Eds.). (2018). *The Cambridge handbook of expertise and expert performance*. Cambridge: Cambridge University Press.

Gawande, A. (2010). *The checklist manifesto*. Profile Books.

George, R. Z., & Bruce, J. B. (Eds.). (2014). *Analyzing intelligence: National security practitioners' perspectives* (2nd ed.). Washington, DC: Georgetown University Press.

Goldstein, D. G., & Gigerenzer, G. (2002). Models of ecological rationality: The recognition heuristic. *Psychological Review, 109*(1), 75.

Gross, S., O'Brien, B., Hu, C., & Kennedy, E. (2014). Rate of false conviction in capital cases. *Proceedings of the National Academy of Sciences, 111*(20), 7230–7235.

Hall, C. C., Ariss, L., & Todorov, A. (2007). The illusion of knowledge: When more information reduces accuracy and increases confidence. *Organizational Behavior and Human Decision Processes, 103*(2), 277–290.

Herzog, M., & Fahle, M. (1997). The role of feedback in learning a Vernier discrimination task. *Vision Research, 37*(15), 2133–2141.

Heuer, R. J. (1999). *Psychology of intelligence analysis* (2nd ed.). Washington, DC: Center for the Study of Intelligence, Central Intelligence Agency. Available at: https://www.cia.gov/library/center-for-the-study-of-intelligence/csi-publications/books-and-monographs/psychology-of-intelligence-analysis/

Heuer, R. J. (2015). *Psychology of intelligence analysis*. Langley: CIA Center for the Study of Intelligence. https://www.cia.gov/library/center-for-the-study-of-intelligence/csi-publications/books-and-monographs/psychology-of-intelligenceanalysis/index.html.

Heuer, R. J., & Pherson, R. H. (2015). *Structured analytic techniques for intelligence analysis* (2nd ed.). Washington, DC: CQ Press.

Kahneman, D. (2011). *Thinking, fast and slow*. New York: Macmillan.

Kahneman, D., Krueger, A. B., Schkade, D., Schwarz, N., & Stone, A. A. (2006). Would you be happier if you were richer? A focusing illusion. *Science, 312*(5782), 1908–1910.

Kruger, J., & Dunning, D. (1999). Unskilled and unaware of it: How difficulties in recognizing one's own incompetence lead to inflated self-assessments. *Journal of Personality and Social Psychology, 77*(6), 1121.

[10] Both women are pregnant.

Loftus, E. F., & Palmer, J. C. (1974). Reconstruction of automobile destruction: An example of the interaction between language and memory. *Journal of Verbal Learning and Verbal Behavior, 13*(5), 585–589.
Loftus, E. F., & Zanni, G. (1975). Eyewitness testimony: The influence of the wording of a question. *Bulletin of the Psychonomic Society, 5*(1), 86–88.
Luft, J., & Ingham, H. (1961). The johari window. *Human Relations Training News, 5*(1), 6–7.
Major, J. (2008). *Communicating with intelligence*. Lanham: Scarecrow Press.
Mercier, H. (2017). Confirmation bias—myside bias. In R. Pohl (Ed.), *Cognitive illusions* (2nd ed.). Abingdon: Routledge.
Miller, G. (1956). The magical number seven, plus or minus two: Some limits on our capacity for processing information. *The Psychological Review, 63*, 81–97.
Nisbett, R. E., Zukier, H., & Lemley, R. E. (1981). The dilution effect: Nondiagnostic information weakens the implications of diagnostic information. *Cognitive Psychology, 13*(2), 248–277.
Nkwake, A. M. (2013). Why are assumptions important? In *Working with assumptions in international development program evaluation* (pp. 93–111). New York: Springer.
Pugh, S. (1991). *Total design: Integrated models for successful product engineering*. New Jersey: Addison-Wesley.
Ross, L., Lepper, M. R., & Hubbard, M. (1975). Perseverance in self-perception and social perception: Biased attributional processes in the debriefing paradigm. *Journal of Personality and Social Psychology, 32*(5), 880–892.
Rossmo, D. K. (2008). *Criminal investigative failures*. London: CRC press.
Safire, W. (2004). *No uncertain terms*. New York: Simon and Schuster.
Simon, H. A. (1956). Rational choice and the structure of the environment. *Psychological Review, 63*(2), 129–138.
Skinner, B. F. (1948). 'Superstition' in the pigeon. *Journal of Experimental Psychology, 38*(2), 168.
Slovic, P. (2018). *Behavioral problems of adhering to a decision policy*. Unpublished manuscript, available at: https://scholarsbank.uoregon.edu/xmlui/handle/1794/23607.
Smith, M. R., & Alpert, G. P. (2007). Explaining police bias: A theory of social conditioning and illusory correlation. *Criminal Justice and Behavior, 34*(10), 1262–1283.
Spencer, K. B., Charbonneau, A. K., & Glaser, J. (2016). Implicit bias and policing. *Social and Personality Psychology Compass, 10*(1), 50–63.
United Nations Office on Drugs and Crime. (2011). *Criminal intelligence manual for analysts*. New York: United Nations. Available at: https://www.unodc.org/documents/organized-crime/Law-Enforcement/Criminal_Intelligence_for_Analysts.pdf.
United States Department of Defense. (2017). *FM34–52 Human intelligence collector operations: FM 2-22.3*. Washington, DC.
White, D., Kemp, R. I., Jenkins, R., Matheson, M., & Burton, A. M. (2014). Passport officers' errors in face matching. *PLoS One, 9*(8), e103510.

Front-End Forensics: An Integrated Forensic Intelligence Model

Max M. Houck

Abstract Forensic science has the ability and capacity to be an active participant in investigations, rather than its traditional passive role. Too often, the evidence is collected, submitted to the laboratory, and the investigators wait for results, creating a lag between when the information is needed and when it is made available. Obstacles to forensic service providers being actively involved in investigations include organizational cultural norms, operational resources, sworn and civilian human resource issues, and media-driven expectations of outcomes. While traditional intelligence analysis is forecasting, forensic science and criminal investigations are reconstructive and this alters which methods can be and how to improve investigations. What has been characterized as forensic intelligence in the literature is largely forensic support to crime analysis. The benefits to using actual forensic intelligence actively in investigations by shifting forensic results, even if preliminary, to the front-end of the criminal justice system, are reduced wrongful arrests and convictions, more efficient use of policing resources, and stronger cases for adjudication. An integrated forensic intelligence model (IFIM) is offered as a roadmap to creating a sustained professional culture of forensic intelligence.

Keywords Intelligence · Forensics · Forecasting · Policing

1 Introduction

Intelligence, as a formal product of information analysis, essentially reduces uncertainty (Lowenthal 2006). Uncertainty comes from asymmetric information, whether it be terrorists secretly plotting an attack or criminals clandestinely surveilling a bank to rob. The state use of intelligence tips the asymmetry of information to the state's favor. The classic view of intelligence comes from a country's clandestinely gathering information about foreign nations; law enforcement adopted this concept for domestic, criminal investigative purposes. One adjunct to the forms of

M. M. Houck (✉)
Forensic Studies & Justice Program, University of South Florida, St. Petersburg, FL, USA
e-mail: maxmhouck@mail.usf.edu

B. Fox et al. (eds.), *Science Informed Policing*, Advanced Sciences and Technologies for Security Applications,
https://doi.org/10.1007/978-3-030-41287-6_8

intelligence used in criminal investigations is forensic science, which uses scientific methodology to reveal relationships between people, places, and things involved in crimes (Houck and Siegel 2015). Forensic intelligence is not a new concept but one that has been misused in the classic sense of intelligence and is under-utilized in criminal investigations. Much of what appears in the literature as forensic intelligence is actually forensic science in support of crime analysis. This chapter explores the basis of intelligence, its development in law enforcement, the state of forensic science provision in the U.S., and the obstacles that need to be overcome to move forensic science from a passive to an active participant in reducing crime through intelligence. An integrated model of forensic intelligence (IFIM) is offered at the end as a roadmap to creating a sustained professional culture of forensic intelligence.

2 Intelligence

Intelligence is different from other governmental functions in that much of what is done is secret and that secrecy seems to run counter to the other functions of government which are public (Lowenthal 2006). And if intelligence were any other governmental function, a definition would probably not be needed--everyone knows what road repairs and garbage collection are. Intelligence is often confused with information and the two are intimately related. **Information** is anything that can be known without regard for how it was discovered or learned. By contrast, **intelligence** is information that meets the expressed needs of government and has been collected, processed, and narrowed specifically to meet these needs. Thus, all intelligence is information but not all information is intelligence (Lowenthal 2006).

Intelligence is produced in a series of steps referred to as the intelligence cycle (Fig. 1). Although approximately accurate for teaching purposes, the true pathway of the intelligence process can be quite complicated (Fig. 2). In any intelligence process, unanticipated issues arise, like additional collection, questions of processing, and changing requirements, among others, that cause additional, layered intelligence processes to occur. As Lowenthal notes, the intelligence process is really "linear, circular, and open-ended all at the same time." (Lowenthal 2006; page 66).

Collection occurs from a range of sources defined by the relevant discipline that specializes in collecting that type of information for intelligence purposes (INT)[1]:

- Human (HUMINT): From a person or people
- Geospatial (GEOINT): From satellite and aerial photography or mapping data
- Measures and signatures (MASINT): From an array of distinctive characteristics of target sources, including electro-optical, nuclear, radar, geophysical, materials, and radiofrequency
- Open Sources (OSINT): From open (unclassified), publicly-available sources.

[1] https://www.dni.gov/index.php/what-we-do/what-is-intelligence

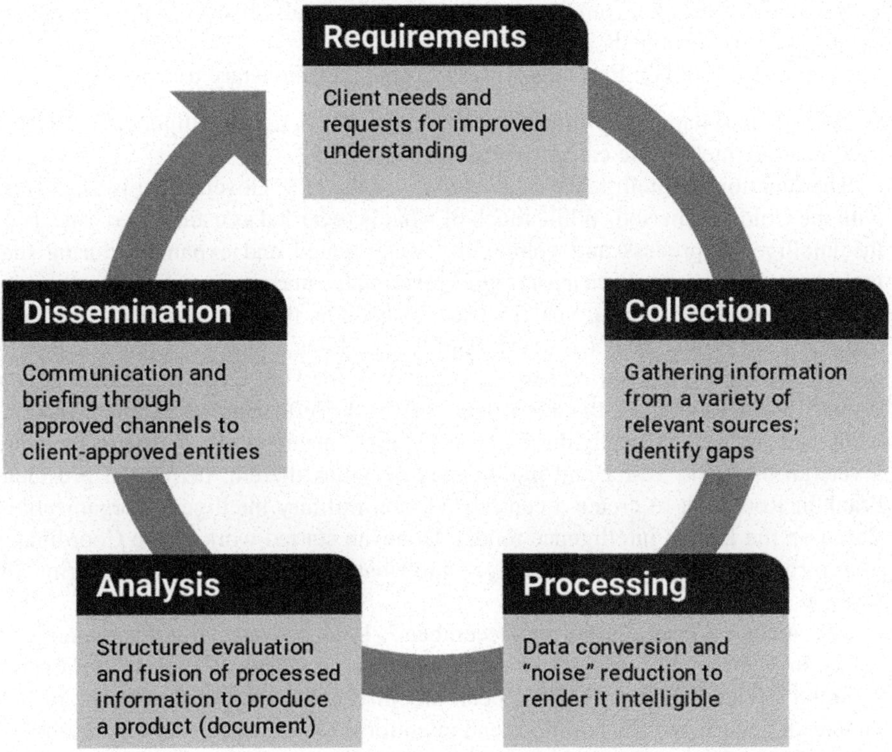

Fig. 1 The Intelligence Cycle: Requirements for intelligence are expressed, the information is collected, it is processed and analyzed (thus making it intelligence), then disseminated to the appropriate entities

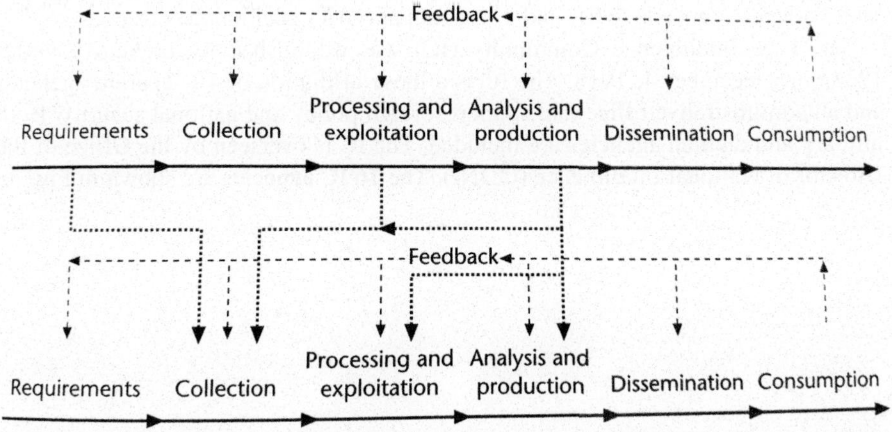

Fig. 2 A more realistic view of the intelligence "cycle." In any intelligence process, unanticipated issues arise, like additional collection, questions of processing, and changing requirements, among others, that cause additional, layered intelligence processes to occur. (Lowenthal 2006; page 67)

- Signals (SIGINT): From the interception of signals, like communications or other electronic signals
- Financial (FININT): From analysis of financial or monetary transactions

Later in this chapter, the concept and category of forensic intelligence (FORINT) as a separate intelligence category will be re-examined.

The function of intelligence as an activity of the U.S. Government is associated with the Cold War period. While much of what is regarded as established aspects of the intelligence process and community was created and expanded during that period, intelligence has been a part of the government since its founding. Intelligence functions originated during war time (as far back as the Revolutionary War[2]) but expanded during the early twentieth century largely as a series of domestic intelligence capabilities; this period saw the creation of the U.S. Department of Justice's Bureau of Investigation, the forerunner of the Federal Bureau of Investigation, whose role expanded greatly during World War II. In early 1941, William Donovan, a veteran of World War I and intelligence activities therein, petitioned President Franklin Roosevelt to create a centralized, non-military intelligence organization. Based on the British intelligence model, Donovan started work as the Coordinator of Information to collect and analyze information that may have a bearing on national security.

The attack on Pearl Harbor on December 7, 1941, revealed significant gaps and lapses in U.S. intelligence efforts. Investigations revealed casual, uncoordinated handling of intelligence that had been sloppily collected. Analysts had grossly underestimated Japan's capabilities and intentions, largely due to American cultural biases. The Office of Strategic Services (OSS) was established in June, 1942 and Donovan was left in charge as the new Director. President Harry Truman issued an executive directive, the National Security Act of 1947, creating a centralized intelligence apparatus, the National Security Council, and a civilian Central Intelligence Agency (CIA) under a Director of Central Intelligence. The Act also laid the boundaries between domestic (FBI) and international (CIA) intelligence efforts.

The U.S. Intelligence Community (IC) was established by Executive Order 12333 on December 4, 1918. The IC is a loose affiliation of 16 separate agencies and an administrative office that support foreign policy and national security. Both military and civilian agencies are included. The IC is overseen by the Office of the Director of National Intelligence (ODNI). The 16 IC agencies are shown in Fig. 3.

---

[2] The Continental Congress created the Committee of Secret Correspondence in 1775 to collect intelligence from people in England, Ireland, and Europe to help in the war effort. https://founders.archives.gov/documents/Franklin/01-22-02-0167

Agency	Parent Agency	Federal Department	Established
Office of Naval Intelligence	United States Navy	Defense	1882
Coast Guard Intelligence	United States Coast Guard	Homeland Security	1915
Bureau of Intelligence and Research	None	State	1945
Central Intelligence Agency	None	Independent agency	1947
Twenty-Fifth Air Force	United States Air Force	Defense	1948
National Security Agency/Central Security Service	None	Defense	1952
Defense Intelligence Agency	None	Defense	1961
National Reconnaissance Office	None	Defense	1961
Intelligence and Security Command	United States Army	Defense	1977
Office of Intelligence and Counterintelligence	None	Energy	1977
Marine Corps Intelligence Activity	United States Marine Corps	Defense	1978
National Geospatial-Intelligence Agency	None	Defense	1996
Office of Terrorism and Financial Intelligence	None	Treasury	2004
Intelligence Branch	Federal Bureau of Investigation	Justice	2005
Office of National Security Intelligence	Drug Enforcement Administration	Justice	2006
Office of Intelligence and Analysis	None	Homeland Security	2007

Fig. 3 The agencies composing the U.S. intelligence community. (*Source*: http://www.dni.gov/index.php/intelligence-community/members-of-the-ic)

3 Intelligence-Led Policing

As with most things military, the diffusion of intelligence from a strictly wartime activity has diffused to civilian government activities, as with the CIA, where it coalesced into its modern post-Cold War structure. From there, intelligence was adopted across a number of disciplines and sectors, including law enforcement. The demand for what is now called intelligence-led policing (ILP) was growing but was cemented by the UK through the adoption of the National Intelligence Model in 2000. The Model shifted from a business model where operations dictated intelligence-gathering priorities to one where criminal intelligence guided operations and policy. The Model concentrated on:

- Targeting offenders
- The management of crime and disorder hotspots
- The investigation of linked series of crimes and incidents, and
- The application of preventative measures.

The Model was later adopted and adapted in many countries, including the United States, Canada, Australia, and New Zealand (Burcher and Whelan 2019).

Because the governmental remit and demands of policing are significantly different from those of statecraft and diplomacy, aspects of the intelligence process were explored and adapted to law enforcement. One of these amendments is the 3-I model proposed by Ratcliffe (2003) (Fig. 4). In Ratcliffe's model, the law enforcement agency engaging in intelligence work should be able to interpret the criminal environment it works in, use that interpretation to influence the decision-makers towards crime reduction strategies, and finally be able to demonstrate the impact those strategies have on the criminal environment. One of the important aspects of

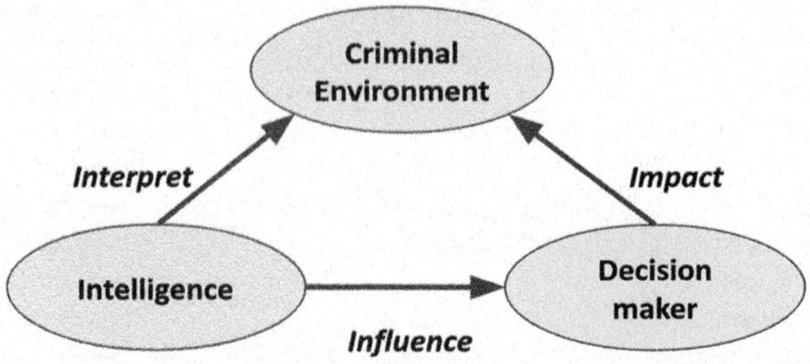

Fig. 4 The 3-I model as developed by Ratcliffe (2006). The law enforcement agency engaging in intelligence work should be able to interpret the criminal environment it works in, influence the decision-makers towards crime reduction strategies, and finally be able to have an impact on the criminal environment

the 3-I model is the demonstrating of impact: Citizens are taxpayers and are arguably a client or at least a stakeholder in funding and supporting which strategies work. The alternative is also true, as Ratcliffe points out, "[t]here is a tendency, however, for what really works in policing to be overshadowed by schemes that have significant public appeal but for which the crime reduction evidence is disputed" (Ratcliffe 2016, page 4). Thus, in Ratcliffe's view, intelligence-led policing strategies should also be evidence-based to support what works and to avoid what does not.

While ILP has been widely adopted conceptually, putting it into actual practice has met with numerous pitfalls and obstacles, including (Burcher and Whelan 2019):

- Data challenges, accuracy, sharing, and overload
- Having systems and software to collect, analyze, and disseminate information
- Lack of ongoing training
- Inconsistencies of agencies' understanding of intelligence, its production, and its use

The boundaries within and between police organizations, both formal and informal, shape how intelligence is viewed, created, used, and supported. Gerringer and Bart (2016, page 324) note further limitations to the effective use of intelligence by law enforcement:

> In many instances, the budgets for law enforcement agencies are too constrained to allow for sufficient intelligence capabilities. Law enforcement intelligence units often cannot analyze collected data because of their limited personnel. Differing federal, state, and local laws and overlapping jurisdictions can inhibit the effective sharing of law enforcement intelligence between the tiers of agencies. Furthermore...organizational and personal jealousies can have negative effects…[t]he inherent secrecy that cloaks intelligence also fosters suspicions of improper behavior by law enforcement and infringements of civil liberties.

Perceived violations legal or moral principles typically provokes a political reaction which often further constrains law enforcement intelligence activities.

Despite these limitations, the use of intelligence by law enforcement has expanded post-September 11. Beyond federal agencies, state and many large city law enforcement agencies have intelligence units or at least capabilities. Joint Terrorism Task Forces (JTTF) combine and coordinate federal, state, and local intelligence capabilities, for example; over 100 JTTFs are distributed across the United States (Gerringer and Bart 2016).

4 Law Enforcement Intelligence

A primary distinction between criminal investigations and traditional intelligence is that investigations are retrospective in nature, working to reconstruct events that have occurred and are in the past, while intelligence is prospective, intending to predict and prevent possible future events. Moreover, as Gerringer and Bart (2016) note, criminal investigations yield evidence suitable for criminal justice purposes, particularly prosecutions, while intelligence produces interpretations of the future from incomplete information about it (Fig. 5). Criminal investigations are intended to be public as a part of the criminal justice system but if this were a mandate for intelligence, it would negate its value. Thus, inherent in the definition of classic intelligence is its covert nature, one that has taken on less-acceptable connotations for law enforcement intelligence activities (Carter 2004).

With the growth of both crime analysis and crime intelligence, it is important to distinguish the two for purposes of this chapter. The difference between crime analysis and law enforcement intelligence is that,

> Crime analysis focuses on analyzing a series of crimes—most notably homicide, assault, robbery, burglary, and auto theft—that have already occurred, with the intent of apprehending the offender(s) and deterring continued criminal acts. Conversely, intelligence analysis assesses diverse types of information that suggest potential criminality—such as suspicious activity reports, tips, and leads—for the purpose of identifying a criminal threat that is typically trans-jurisdictional in nature, with the purpose of intervening to stop the threat (Global Advisory Committee 2013).

Both disciplines use a scientific approach to problem solving but the analytic techniques used to solve them differ because each has different goals. Crime analysis examines crimes that have already occurred and looks for trends, patterns, and

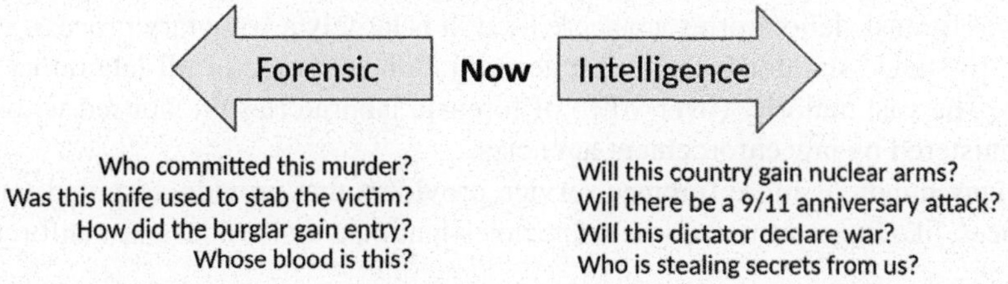

Fig. 5 The critical difference between criminal investigations and classic intelligence is that investigations are retrospective, focusing on past events, while intelligence is prospective, focusing on likely future events

consistencies to understand patterns of crime. Crime analysis provides important information to policy and decision makers about crime in their jurisdiction. In doing so, they use a wider variety of methods from a range of disciplines (sociology, geography, data analysis, psychology, among many others). Conversely, intelligence analysis primarily focuses on the threat environment (related to Ratcliffe's "criminal environment"). Criminal enterprises (including gangs) and their ways of operating, demand for and distribution of illegal goods, cross-jurisdictional crime and threats to the safety and well-being of citizens are all the type of activities that are identified and evaluated through intelligence analysis. The goal of intelligence analysis is to offer approaches to intervene and mitigate or remove the threats at the strategic and tactical levels. Intelligence analysis uses more specific techniques to analyze raw information and convert it to intelligence (see Houck, "Improving Investigations with Structured Analytic Techniques," elsewhere in this publication).

The distinctions described in this Section are important to understanding the growth within and current definitions of forensic intelligence in law enforcement agencies.

5 Forensic Service Providers

Forensic science uses scientific methodology to demonstrate relationships between people, places, and things involved in criminal activity. It does this through the examination and analysis of the evidence left behind, so-called traces, that are indicators of the past criminal activity (Houck and Siegel 2015). In the United States, there are just over 400 forensic science laboratories at the local, state, and federal branches of government that provide a range of analytical and technical (Durose et al. 2016). It is estimated that in 2014 these facilities received 3.8 million request for services (down from 4 million in 2009); about 75% of these requests were for the analysis of controlled substances, biological samples to be submitted to a DNA database, and toxicology. In 2014, forensic science laboratories had a combined estimated backlog of 570,100 cases that had yet to be completed, down from nearly 900,000 in 2009. Not all forensic laboratories provide every type of analysis and whether they do depends on budgets, client requests, and crime rates (Fig. 6). Nearly 90% of forensic laboratories are accredited; it is largely a voluntary process, with at least 10 states[3] statutorily requiring accreditation for some or all laboratory functions. The vast majority (over 90%) of forensic laboratories are housed within and administered by law enforcement agencies.

Many hundreds more forensic service providers that provide a limited range of services, like fingerprint or crime scene investigation units, reside in law enforcement

[3] California, Hawaii, Indiana, Maryland, Missouri, Nebraska, New York, North Carolina, Oklahoma and Texas. Almost all state forensic laboratories (99%) are accredited.

Fig. 6 Services provided by publicly funded forensic laboratories, by type of jurisdiction, 2014. (Durose et al. 2016)

Forensic function	Federal	State	County	Municipal
Controlled substances	55	87	86	71
Crime scene	42	48	51	75
Digital evidence	54	10	20	36
Firearms/toolmarks	27	58	60	58
Forensic biology casework	27	71	68	42
Database samples	12	25	9	4
Impressions	26	43	46	35
Latent prints	67	53	62	88
Questioned documents	34	12	12	14
Toxicology	9	48	52	36
Trace evidence	57	53	50	29

agencies. An exact count of these units is difficult, given the roughly 18,000 law enforcement agencies in the U.S. with some number of those providing a forensic function. Only a few of these units are accredited (Childs et al. 2009). Private forensic laboratories exist but typically specialize in a limited, specialized range of analyses, like DNA or toxicology.

Not all law enforcement agencies have an integrated, cooperative relationship with their forensic divisions. The reasons for that are discussed next.

5.1 Forensic Science Evolved Within a Policing Context

5.1.1 A Little History

In 1910, Edmund Locard established the first forensic laboratory in the world in Lyon, France; it would be another 13 years before the first forensic laboratory would be created in the U.S. by the Los Angeles Police Department. In 1930, the Federal Bureau of Investigation (FBI) established its Bureau of Identification (fingerprints), followed 2 years later with the creation of its own forensic laboratory (November 24, 1932). Other types of operational forensic science laboratories were instituted in the U.S. after this and later expanded due to funding from the Office of Law Enforcement Assistance and its successor, the US Law Enforcement Assistance Administration (LEAA) to improve law enforcement's access to forensic testing in the 1960s and 1970s (Benson et al. 1968; Parker and Peterson 1970).

The translation of the forensic laboratory from Europe to the U.S. resulted in the laboratory functions shifting from an independent university setting to a subservient administrative section within a law enforcement agency (Fosdick 1969; Wilson 1975). The autonomy of the university, particularly in the US with the notion of academic freedom, was largely gone and the purpose of the scientist was now to respond to the police's needs:

> Without a doubt, the laboratory, as it exists in the United States, is an appendage of a quasi-military operation of an enforcement agency. As in the military, the laboratory technician in the quasi-military operation is subordinate to the administration, which is usually not technically trained. The technician, therefore, does not have the freedom of decision nor the opportunity for research that would exist if he were a dedicated, well-trained scientist acting as a civilian in the proper framework (Wilson 1975, pages 99–100).

The forensic laboratory, from its inception, was seen as desirable, "cutting edge" science used to enhance older investigative methods (the "needle in the haystack" method of human intelligence and shoe leather). The desirability of having a laboratory, aggravated by its media status as a new method of catching criminals (one earlier version of a "*CSI* Effect" (Houck 2006)), resulted in many laboratories being hastily conceived and brought into being, resulting in numerous unanticipated problems, such as the professional roles and independence of scientists, salary differentials between scientists and officers, oversight of scientists by non-scientific sworn personnel, and a misunderstanding of technicians vs. scientists (Fong 1975). The proliferation of new laboratories stressed the resources of police agencies along with the increased and novel demands, administrative as well as financial, that science entailed: Capital equipment, training, and personnel. These pressures further subsumed the scientists' work under that of the investigator who now had to juggle not only his own cases but manage the scientists and their increasingly complex scientific techniques (Johnson et al. 2008).

Despite the modernity of "scientific detection" and any perceived benefits accrued, the laboratory was also an "incomprehensible" novelty to the non-scientific police:

> Some of these [law enforcement] agencies which are so eager to have a laboratory have demonstrated to the author's satisfaction that they don't even know what a laboratory is for. Even worse, they have little or no conception of the proper use of a laboratory (Fong 1975, page 100).

For example, one early police science text noted that,

> In general, American detectives do not place much weight upon the application of scientific principles to the solution of the crimes which they are called upon to investigate. There is a reason for this: They place more stress on their lines of information and their acquaintance with criminals and criminal methods (Perkins 1942, page 39).

Competing for resources within a police agency made it harder for forensic science to advance:

> The forensic scientist is conservative in his outlook, partly because over the years he has had to acquire his methods slowly and painfully: I have deliberately used the word painfully because the acquisition has been accomplished against the background of growing work load, shortage of money for equipment, and in the face of some skepticism of the value of scientific evidence from the legal profession (Williams 1975, page 127).

Technology advances, culture lags. Although forensic science has the current status of being a high-tech, cutting edge scientific crime fighter, the historical attitudes and organizational boundaries still exist that prevent forensic science from being fully-integrated into criminal investigations. Moreover, the view of laboratories as passive resources rather than active participants has driven resources to reduce turnaround times and costs and increase productivity. This has lead to the perverse situation where for every unit reduction in backlogged cases, demand increases by more than that unit amount (Speaker 2013). Success in providing resources drives increased demand against already strained resources. So, the question is not whether law enforcement uses forensic services, it is why do they not use them earlier in the investigative process?

6 Political Asymmetry

The vast majority of forensic service providers are governmental and operate in a political environment. Politics can, for better or worse, trump science when it comes to operational decisions (policy, financial, perceptual). Therefore, forensic service providers may be subject to adverse outcomes because of political decisions.

As agents of the state, governmental forensic service providers are dependent on their jurisdictional clients (police, prosecutors) for their raison d'être. In this context, dependence means that one entity (the dominant one) can expand and be (more or less) self-sustaining while others (the dependent ones) can do this only as a reflection of that expansion. This subsidiary expansion can have a positive or negative outcome for the dependent entity. Dependence means an asymmetrical power relationship between the two, with the subordinate entity having little or no ability to grow, change, or expand; without significant change, the subordinate is relegated to economic and political dependence. "In essence, if the rules of the game are biased against you, then you have two choices: to continue to play by the rules and thus continue to be exploited or to rewrite the rules in a way that does not leave you at an unfair disadvantage" (Ishiyama and Breuning 2010, page 330).

Suggestions to rewrite the rules came in the form of the National Academy of Sciences's recommendation in 2009 that forensic service providers be administratively or financially independent of law enforcement-based parent agencies:

> Scientific...assessment conducted in forensic investigations should be independent of law enforcement efforts either to prosecute criminal suspects or even to determine whether a criminal act has indeed been committed. Administratively, this means that forensic scientists should function independently of law enforcement administrators. The best science is conducted in a scientific setting as opposed to a law enforcement setting. Because forensic scientists often are driven in their work by a need to answer a particular question related to the issues of a particular case, they sometimes face pressure to sacrifice appropriate methodology for the sake of expediency. (National Academy of Sciences 2009, pages 23–24).

Suggesting is one thing, success is another. Several jurisdictions have attempted independent forensic agencies with varying success. Power imbalances still exist in these so-called experiments, leading to adverse selection and moral hazards. Adverse selection occurs when buyers and sellers have access to different information, leading to undesired results. If a bank set one price for all checking account customers, it will lose money on both the low-balance and high activity customers.

Likewise, if police and attorneys demand that all cases are a "priority", then none can be: All suffer equally. Moral hazards result when one party takes risks because they will not incur the potential cost of taking it. For example, when a prosecutor demands that evidence be tested even though experts from the dependent laboratory explain that the results will not provide useful information (like testing a handgun for the DNA of the person who had the gun in their possession at arrest), the prosecutor is incurring a risk of resources on the laboratory. Unnecessary work needlessly consumes resources, wastes time, and delays the working of meaningful evidence. Without realizing it, the prosecutor is also incurring an opportunity cost by slowing down the lab's responsiveness on the next case.

Asymmetric information begins a downward spiral, with relational and operational imbalances leading to adverse selection and moral hazards. Ultimately, this may result in market failure, where the allocation of resources is not inefficient. If other conceivable outcomes exist where a situation can be improved without another being made worse, then there is market failure. Multiple scenarios for forensic science provision are possible which would improve the quality and delivery of the science without adversely affecting justice outcomes. Arguably, the forensic industry is approaching or even experiencing a type of market failure. A potential example of forensic market failure can be seen in the crisis in forensic science provision in the United Kingdom leading to an inquiry by its House of Lords.[4]

This asymmetry places forensic services in a passive, receptive space: Submit the evidence and wait for the reports. The transactional relationship combined with the organizational priority of policing over science means that forensic science is seen as a operational cul-de-sac rather than an equal partner in the investigative process. And yet, the use of forensic intelligence, or at least that phrase, expands unabated (Milne 2012). Why is forensic science not used as an active component in criminal investigations? Ross (2015) notes three barriers to greater engagement in the creation and use of forensic intelligence: Organizational, individual, and collective. The organizational barriers, discussed briefly above, mean that isolated individuals (forensic experts) are only brought to the foreground on complex or high profile cases. This leads to individual barriers, where potential participants do not see the value they could bring to the criminal intelligence sharing process, overestimate the work involved in full participation, and ultimately do not know how to contribute to the process (US Department of Justice, et al. 2008). Finally, the collective barrier consists of the cumulative effects of the other two resulting in intelligence-led information is never actually transformed into intelligence products (IACP 2002). Ross (2015) further lists additional barriers to specific to forensic agencies:

- Intelligence analysis and reporting does not comply with traditional quality assurance demanded by accreditation,
- A reticence among forensic scientists to disseminate 'soft' information or preliminary reports for fear it will used against them in court,

[4] https://www.parliament.uk/forensic-science-lords-inquiry

- Forensic laboratories lack of awareness about the needs of partners in policing,
- Many forensic agencies lack IT systems that are fit for intelligence purposes, mostly connectivity and integration.

These barriers would need to be addressed in any comprehensive model that supports the integration of forensic information into the criminal intelligence process.

7 Forensic Intelligence

The literature on forensic intelligence extols its utility in solving linked cases involving controlled substances (Morelatto et al. 2013), fingerprints (Ross 2015), shoeprints (Ribaux et al. 2006)), and DNA (Walsh et al. 2002). The majority of literature emphasizes the use of databases to connect people or crimes through one or more evidence types (Ribaux and Wright 2014). While computerized, the methods ultimately are based around Link Analysis or Relationship Matrix Analysis (Heuer and Pherson 2015). Combined with other domains of information (like geographic mapping, political boundaries, and crime reporting, for example), the product of scientific analysis of collected evidence helps to link otherwise apparently unrelated cases. But is this forensic *intelligence* despite that being what it is called?

7.1 Scale, Scope, and Support of Law Enforcement Intelligence

7.1.1 Scale

The scale of law enforcement intelligence crosses the usual organizational triad of strategic, operational, and tactical.

7.1.1.1 Strategic

Strategic plans involve the entire organization and its mission; they involve leadership. A strategic plan answers questions like, what do we do, for whom do we do it, and how do we excel. The strategic level always encompases at least the largest jurisdictional area involved and typically incorporates a view beyond that (state, national, or transnational).

7.1.1.2 Operational

Operational plans are made by frontline managers and focus on the specific procedures, processes, and methods that occur within the lowest levels of the organization. It answers questions such as, where are we now, where do we want to be, how do we get there, and how do we measure our progress? Therefore, operations equates to the utilization of resources, that, is, management. The combined actions and decisions of operational units influence this view. Information about localized property crime, for instance, is of operational intelligence value. The distribution of illegal drug sales is of operational intelligence value. For law enforcement, the operational level equates to the state or region.

7.1.1.3 Tactical

Tactical plans translate strategy and operations into specific plans relevant to a distinct area of the organization; they are the application of operational demands to meet strategic goals. Tactical plans are concerned with the responsibility and functionality of lower-level departments to fulfill their parts of the strategic plan. The individual officer on patrol has an effect on tactical intelligence; they may collect raw information or produce lightly-analyzed intelligence. How gang members operate locally is of tactical intelligence value. Tactical intelligence from one officer can be combined with information or intelligence from several other officers to begin developing an operational view.

7.1.2 Scope

For example, activities of a local gang's members in a city is of tactical intelligence value. If that gang cooperate with other gangs in the region (county, state, etc.) that would be of operational intelligence value. If those gangs then have interactions with affiliates across multiple states, that would be of strategic intelligence value. Tactical information and intelligence feeds operational intelligence which informs strategic plans. Likewise, unless the subordinate operational levels are appreciated and understood, the strategic plan would be misinformed and likely to be unsuccessful.

7.1.3 Definitions

The distinctions made between classic intelligence and crime mapping are of assistance in answering this question. If classic intelligence focuses on the threat environment and Ratcliffe's criminal environment in criminal intelligence is a useful analog, then forensic intelligence would focus on forensic reports that help to reveal the operation of criminal enterprises, identification of illegal goods, and work trans-jurisdictionally to intervene and mitigate or remove the threats at the strategic and tactical levels. Forensic intelligence would be case-based, then, relating to individual or serial instances of specific crimes (operational to tactical).

Conversely, forensic science can add value to crime analysis by connecting scientifically-revealed details that demonstrate trends, patterns, and consistencies to understand occurrences of crime. Forensic science would then be one more discipline folded into the larger, policy-related strategic goals of crime analysis (strategic to operational).

7.2 Forensic Intelligence Is Not Crime Analysis

What has been predominantly discussed in the literature about the use of forensic results in criminal investigations is supporting crime analysis, not conducting intelligence (for example, see Morelato et al. 2013, 2014). This does not lessen the value it brings to the solution of crimes and the agencies who solve them but is a reminder of the slipperiness of terminology when it crosses professional borders. For example, Ribaux and co-authors, from their forensic viewpoint state, "The collection of data is the object of great attention in typical intelligence processes: It must be selective, timely, and planned according to relevancy and accessibility of data, as well as the availability of resources" (2010, page). This ignores several tenets of classic intelligence collection:

> The collection process encompasses the management of various activities, including developing collection guidelines that ensure optimal use of available intelligence resources. **Intelligence collection requirements are developed to meet the needs of potential consumers.** Based upon identified intelligence, requirements collection activities are given specific taskings to collect information. **These taskings are generally redundant and may use a number of different intelligence disciplines for collection activities.** Tasking redundancy compensates for the potential loss or failure of a collection asset. It ensures that the failure of a collection asset is compensated for by duplicate or different assets capable of answering the collection need. **The use of different types of collection systems contributes to redundancy.** It also allows the collection of different types of information that can be used to confirm or disprove potential assessments" (FAS 1996, Section 2; emphasis added).

This comparison is not to nitpick but to show that much of the current thinking on forensic intelligence has been developed without regard to the classic intelligence framework and its professional community. Simply calling work "forensic intelligence" does not make it proper intelligence work. What has been published in the literature is more properly crime analysis and has contributed a great deal to moving forensic science forward in the investigative process.

Forensic science can improve criminal investigations at the front end but, as noted by Ross, usually only in high profile or complex cases. The work done in the UNABOMB case, using forensic results of individual bombings to divine signatures of Kaczynski, is an example among many others (Freeman et al. 2018). Lesser known examples are abundant but not couched in the domain of intelligence. For example, evidence from the sexual assault and murder of a young girl revealed the presence of small fiber nubs, a by-product of the fabric finishing process called singeing, which uses heat to smooth the surface of a fabric. Traditional police work led to several potential suspects who worked at a fast food restaurant. Work uniforms were collected from those employees who were off-shift and unaccounted for at the time of the crime. Forensic fiber analysis revealed the presence of nubs on only one employee's shirt; the company had recently changed uniform suppliers and the new one did not purchase fabric from a source that used singeing (Deedrick 2001) (Fig. 7).

Another example using fibers involves a series of sexual assaults of children in rural Michigan and Ohio. The perpetrator would find a child on a secluded location using a gun would command them into his vehicle. Once inside, the perpetrator would have them duct tape their mouths closed and cover them with a blue blanket-type fabric to hide them from view. Once assaulted, the victims were released and the bindings recovered. Blue polypropylene fibers were found in common between a number of the victims. One witness described a white Oldsmobile vehicle near one scene. The laboratory contacted General Motors and compared the blue fibers to the master standards for textiles installed in their vehicles. The color and fiber types narrowed the possible sources down to four model types from the late 1980s (the crimes occurred from 1988 to 1991). A subsequent victim saw some of the vehicle interior and described an emergency brake cover between the front seats with a herringbone pattern on it. Consultation with the car manufacturer revealed that this brake style as used from 1986 to 1988 in the Calais model. Authorities, in cooperation with the car manufacturer, eventually narrowed down the number of

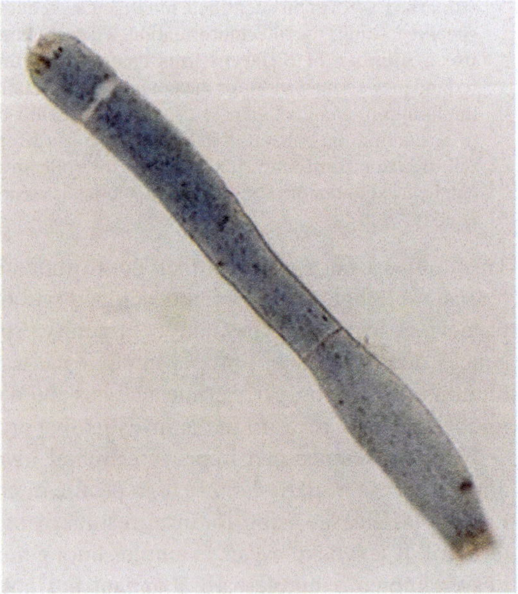

Fig. 7 Singeing is a process used to finish fabric surfaces to smooth them. The fabric is run under a heat source that melts the synthetic fibers and reduces any stray fibers that stick up. The unintended result are nubs, small melted bits of the fibers that stick to the shirt and can transfer to other surfaces, like from a perpetrator to a victim. (Deedrick 2001)

1986 to 1988 white Oldsmobile Calais with blue interiors sold in Michigan and Ohio to six. Surveillance and subsequent laboratory analyses led to the arrest and conviction of the perpetrator (Michaud 2001).

Examples such as these can and have filled books (Houck 2001, 2004). The issue is not that these examples exist but that the approach of moving forensic intelligence to the forefront of criminal investigations has not been codified and routinely integrated for the reasons previously mentioned. This means that an untold amount of potentially useful and powerful intelligence is going unused for lack of a model to bring it into the criminal investigative fold.

8 Conclusions

As Ross (2015) states, "[f]orensic intelligence should not stand on its own. It should become an integral part of the overall investigative tool and intelligence-led strategies." This means that forensic intelligence needs to be defined and addressed with three spheres:

- Intelligence as a **process**: The means by which forensic information required and requested, collected, analyzed, and disseminated.
- Intelligence as a **product**: The outcome of these processes.
- Intelligence as an **organization**: Units that carry out these functions.

To do this within an accepted model, Ross outlined (Ross 2015) the elements of a forensic intelligence model:

- Timeliness (24 h to 5 days)
- Interim reporting ("soft" results)
- Integration of forensic services (best results first; redundancy as necessary)
- Aware and engaged practitioners (creating an intelligence culture in forensic science)
- Informed analysis (professionalization of the forensic intelligence culture)
- Systems enabled information technology (improve flow of communication)
- End-user (client) informed (understanding what the products provided mean)
- Engaged partners and trusted relationships (feedback and reinforcement of value added)
- Legal validation (working within legal framework)
- Re-evaluation (feedback on process, products, and organization)
- Research and development (advance, correct, and improve methods employed) (for a more developed view of research, see Legrand and Vogel 2014)

This list comports with the intelligence community's recognition post-9/11 of the need for collaborative interagency intelligence production. Guidetti (2009) lists the following:

- Clarify mission, roles, and responsibilities,

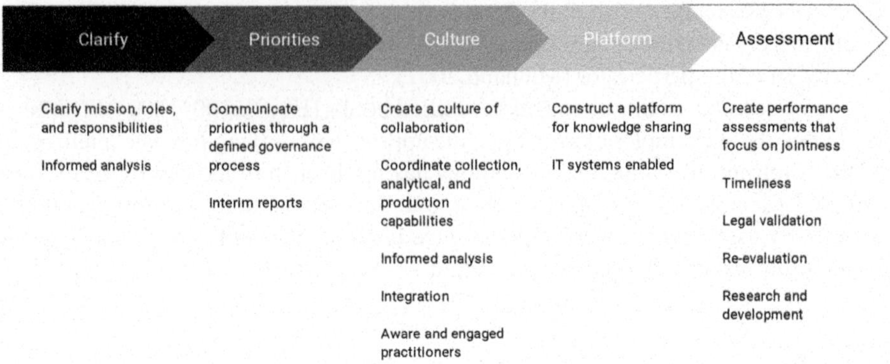

Fig. 8 An integrated forensic intelligence model (IFIM) based on Guidetti (2009) and Ross (2015)

- Communicate priorities through a defined governance process,
- Create a culture of collaboration,
- Construct a platform for knowledge sharing,
- Coordinate collection, analytical, and production capabilities, and
- Create performance assessments that focus on jointness.

Creating a collaborative approach yields "exponential" benefits, according to Guidetti, because there "are more tools in the analyst's arsenal to draw from, while the stakeholder and constituent base amplify. This is significant for intelligence entities that are usually understaffed to support wide-ranging missions" (Guidetti 2009, page 224). Guidetti's six recommendations can be considered the strategic view for integrating forensic intelligence into criminal investigations while Ross' list drives the operational view of what resources are needed. Combining these two levels of views and needs results in an integrated forensic intelligence model (IFIM) (Fig. 8) that provides the broad strokes and next steps to working towards a forensic intelligence community that would be recognized in the larger scheme of the intelligence community.

References

Benson, W., Stacy, J., & Whorley, M. (Eds.). (1968). *Systems analysis of criminalistics and forensic science*. Washington, DC: Law Enforcement Assistance Administration.

Burcher, M., & Whelan, C. (2019). Intelligence-led policing in practice: Reflections from intelligence analysts. *Police Quarterly, 22*(2), 139–160.

Carter, D. L. (2004). *Law enforcement intelligence: A guide for state, local and tribal law*. Washington, DC: Community Oriented Policing Services, U.S. Department of Justice.

Childs, R. A., Witt, T. S., & Nur-tegin, K. (2009). Survey of forensic service providers. *Forensic Science Policy and Management, 1*(1), 49–56.

Deedrick, D. (2001). Fabric processing and "nubs". In M. Houck (Ed.), *Mute witnesses: Trace evidence analysis* (pp. 1–20). New York: Academic Press.

Durose, M. R., Burch, A. M., Walsh, K., & Tiry, E. (2016). *Publicly funded forensic crime laboratories: Resources and services, 2014*. Washington, DC: Bureau of Justice Statistics, U.S. Department of Justice.

Federation of American Scientists (FAS). (1996). *Operations security intelligence threat handbook*. Interagency OPSEC Support Staff. fas.org/irp/nsa/ioss/threat96/index.html

Fong, W. (1975). Criminalistics and the prosecutor. In J. L. Peterson (Ed.), *Forensic science: Scientific investigation in criminal justice* (pp. 371–394). New York: AMS Press, Inc.

Fosdick, R. B. (1969). *American police systems*. Patterson Smith Reprint of a Century Company Work.

Freeman, J., Turchie, T., & Noel, D. (2018). *UNABOMBER: How the FBI broke its own rules to capture the terrorist Ted Kaczynski*. New York: History Publishing Company.

Gerringer, A., & Bart, J. (2016). Law Enforcement Intelligence. In P. Oleson (Ed.), *AFIO's guide to the study of intelligence* (pp. 321–326). Falls Church: Association of Former Intelligence Officers.

Global Advisory Committee. (2013). *National Criminal Intelligence sharing plan: Building a national capability for effective criminal intelligence development and the nationwide sharing of intelligence and information*. Washington, DC: Bureau of Justice Statistics, U.S. Department of Justice.

Guidetti, R. (2009). Collaborative intelligence production. In J. Ratcliffe (Ed.), *Strategic thinking in criminal intelligence* (pp. 222–234). Sydney: The Federation Press.

Heuer, R. J., & Pherson, R. H. (2015). *Structured techniques for intelligence analysis*.

Houck, M. (Ed.). (2001). *Mute witnesses: Trace evidence analysis*. London: Academic Press.

Houck, M. (Ed.). (2004). *Trace evidence analysis: More cases in mute witnesses*. London: Academic Press.

Houck, M. (2006). CSI: Reality. *Scientific American, 295*(1), 84–89.

Houck, M., & Siegel, J. (2015). *Fundamentals of forensic science*. London: Academic Press.

International Association of Chiefs of Police and US Department of Homeland Security. (2002) *Criminal intelligence sharing: A national plan for intelligence-led policing at the local, state and federal levels*. https://it.ojp.gov/documents/National_Criminal_Intelligence_Sharing_Plan.pdf

Ishiyama, J., & Breuning, M. (2010). *21st century political science: A reference*. London: Sage.

Johnson, H. A., Wolfe, N. T., & Jones, M. (2008). *History of criminal justice* (4th ed.). Newark: Matthew Bender.

Legrand, T., & Vogel, L. (2014). The landscape of forensic intelligence research. *Australian Journal of Forensic Sciences, 47*(1), 16–26.

Lowenthal, M. (2006). *Intelligence: From secrets to policy* (3rd ed.). Washington, DC: CQ Press.

Michaud, A. (2001). Trace evidence as investigative lead value. In M. Houck (Ed.), *Mute witnesses: Trace evidence analysis* (pp. 49–68). London: Academic Press.

Milne, R. (2012). *Forensic intelligence*. Boca Raton: CRC Press.

Morelato, M., Beavis, A., Tahtouh, M., Ribaux, O., Kirkbride, P., & Roux, C. (2013). The use of forensic case data in intelligence-led policing: the example of drug profiling. *Forensic Science International, 226*(1–3), 1–9.

Morelato, M., Baechler, S., Ribaux, O., Beavis, A., Tahtouh, M., Kirkbride, P., et al. (2014). Forensic intelligence framework—Part I: Induction of a transversal model by comparing illicit drugs and false identity documents monitoring. *Forensic Science International, 236*, 181–190.

National Academy of Sciences National Research Council. (2009). *Strengthening forensic science in the U.S.: A path forward*. Washington, DC: National Academy of Sciences National Research Council.

Parker, B., & Peterson, J. (1970). *Physical evidence utilization in the administration of criminal justice*. Berkeley: School of Criminology, University of California.

Perkins, R. (1942). *Elements of police science*. Chicago: The Foundation Press.

Ratcliffe, J. (2003). Intelligence-led policing. In *Trends and issues in crime and criminal justice* (Vol. 248). Canberra: Australian Institute of Criminology.

Ratcliffe, J. H. (2016). *Intelligence-led policing*. Routledge.

Ribaux, O., & Wright, B. T. (2014). Expanding forensic science through forensic intelligence. *Science & Justice, 54*(6), 494–501.

Ribaux, O., Walsh, S. J., & Margot, P. (2006). The contribution of forensic science to crime analysis and investigation: Forensic intelligence. *Forensic Science International, 156*(2–3), 171–181.

Ribaux, O., Baylon, A., Roux, C., Delémont, O., Lock, E., Zingg, C., & Margot, P. (2010). Intelligence-led crime scene processing. Part I: Forensic intelligence. *Forensic Science International, 195*(1-3), 10–16.

Ross, A. (2015). Elements of a forensic intelligence model. *Australian Journal of Forensic Sciences, 47*(1), 8–15.

Speaker, P. J. (2013). Forensic science service provider models: data-driven support for better delivery options. *Australian Journal of Forensic Sciences, 45*(4), 398–406.

U.S. Department of Justice, International Association of Chiefs of Police, & US Department of Homeland Security. (2008). *National summit on intelligence: Gathering, sharing, analysis, and use after 9–11*. www.theiacp.org/resources/national-summit-on-intelligence-gathering-sharing-analysis-and-use-after-9-11

Walsh, S., Moss, D., Kliem, C., & Vintiner, G. (2002). The collation of forensic DNA case data into a multi-dimensional intelligence database. *Science & Justice, 42*(4), 205–214.

Williams, R. (1975). In J. L. Peterson (Ed.), *Forensic science–The present and the future, in Forensic science: Scientific investigation in criminal justice* (pp. 126–148). AMS Press, Inc.: New York.

Wilson, C. (1975). Crime detection laboratories in the United States. In J. L. Peterson (Ed.), *Forensic science: Scientific investigations in criminal justice* (pp. 96–108). New York: AMS Press, Inc.

Missing Persons and Runaway Youth: The Role of Social Media as an Alert System and Crime Control Tool

Michelle N. Jeanis

Abstract Missing persons cases have a unique relationship with the media, where increased exposure is said to be a key factor in the recovery of victims and closure of cases. In an attempt to increase exposure for certain types of missing persons cases, news media outlets share case content and alert systems were created; however, the current systems in place have notable limitations. In an attempt to adapt to the proliferation of social media in the lives of everyday Americans, law enforcement agencies have begun to rely on social media as a crime control tool and as a means to raise awareness of missing persons cases. This chapter serves as an overview of the current academic standing of social media as a mechanism of crime control, with a specific focus on how law enforcement can optimize their use of social media in the form of a missing persons outreach tool.

Keywords Missing persons · Runaway youth · Media and crime · Social media

1 Missing Persons Cases

Currently, there are over 88,000 active missing persons cases in the United States (NCIC 2017). The most recent statistics report 651,226 new missing persons cases added to the National Crime Information Center's (NCIC) national database, which is a slight increase over previous years (NCIC 2017). Missing persons cases consist of an individual whose whereabouts are unknown due to one of the following reasons: "physical or mental disability, circumstances that indicate the possibility of danger, indication that the individual did not leave on their own volition, individuals missing during natural disasters, there is concern for the individual's safety, or the individuals is under the age of 21" (NCIC 2017). Fortunately, the majority missing persons cases are resolved quickly, where of those 651,226 new cases, only 63,515

M. N. Jeanis (✉)
University of Louisiana at Lafayette, Lafayette, LA, USA
e-mail: mjeanis@louisiana.edu

B. Fox et al. (eds.), *Science Informed Policing*, Advanced Sciences and Technologies for Security Applications,
https://doi.org/10.1007/978-3-030-41287-6_9

remained open at the end of 2017 (NCIC 2017). Many of the missing persons cases in the United States consist of cases involving children and those under the age of 21, with the typical case consisting of a youth runaway (NCIC 2017).

Missing persons cases are often time consuming for law enforcement at both the local and federal level and can involve costly investigative methods (Fyfe et al. 2015; Greene and Pakes 2014). Although limited in statistics, research suggests that investigation of missing persons cases, especially those concerning stranger abducted children or high profile cases can become very costly and complicated (Fyfe et al. 2015). These cases are also often complex, involving cross-jurisdictional collaboration, civilian volunteers, non-profit organizations, and sometimes federal agency involvement (Fyfe et al. 2015; Morewitz and Colls 2016; The Commission on Peace Officer Standards and Training 2011). Additional pressures from the media and community, as well as the time sensitive nature of these investigations (the majority of stranger child kidnappings result in victim death within the first 3 h), make these investigations especially taxing on law enforcement agents (Biehal et al. 2003; Fyfe et al. 2015; Hickey 2016). Other types of missing persons cases can also pose a significant strain on law enforcement agencies. Long term missing children or adult cases can require investigation by law enforcement for decades. In addition, due to the cyclical nature and prevalence of youth runaway cases, law enforcement agents are often inundated with the search and recovery of missing youth (Whitbeck and Holt 1999; Windle 1988). Furthermore, youth runaways are often at risk for exploitation and delinquency during a runaway event, which further strains law enforcement agencies (Chen et al. 2007; The National Center of Missing and Exploited Children 2015; Whitbeck and Hoyt 1999).

2 The Role of News Media

No matter the type of missing persons case at hand, these types of cases often consist of features that do not typically accompany the average crime investigation (Fyfe et al. 2015). For example, time is especially critical for stranger involved kidnappings and immediate police response is integral to the safe recovery of the victim (Hickey 2016). In regards to runaway youth case investigation, law enforcement not only search for the missing child, but also work to identify family dynamics that may be dangerous to the youth/motivate runaway behavior and also attempt to protect the youth from the dangers of the runaway experience (Chen et al. 2007). Missing persons cases involving individuals with a developmental or mental impairment can also require additional investigative techniques from law enforcement agencies due to the increase likelihood of incapacitation, as the victim is limited in the ability to aid in their safe return to the community (Biehal et al. 2003). Perhaps the most unique feature of missing persons cases involves the reliance on media outlets to share information with the public regarding these types of cases (Fyfe et al. 2015). Law enforcement often need to relay information regarding the missing person to the public, so that they can aid in the search efforts for the victim, and sometimes the offender as well (Fyfe et al. 2015). Although yet to be assessed empirically, law enforcement and practitioners often state that public awareness of a missing persons case greatly increases the likelihood of recovery of the victim (Fyfe et al. 2015).

Unfortunately, not all missing persons cases are treated in the same way by the news media (Jeanis and Powers 2017; Min and Feaster 2010; Moscowitz and Duvall 2011; Simmons and Woods 2015; Taylor et al. 2013). For example, research suggests that cases involving minorities, particularity African Americans, are significantly less likely to receive media attention, compared to missing White individuals (Jeanis and Powers 2017; Newiss 2005; Simmons and Woods 2015). This finding has also been supported for the Indigenous population as well, where Indigenous women are significantly less likely to receive news media attention than White women (Gilchrist 2010). In addition, missing persons cases involving missing men are also less likely to receive media attention, comparted to cases involving missing women (Gilchrist 2010; Jeanis and Powers 2017; Taylor et al. 2013). Disparity by age and case circumstances have also been established, where younger victims are more likely to receive media attention (compared to adults) and stranger kidnappings are more likely to receive media attention than familial abductions (Moscowitz and Duvall 2011; Min and Feaster 2010; Taylor et al. 2013). This established disparity is especially concerning due to the belief that increased news media attention is related to the increased likelihood of victim recovery (Fyfe et al. 2015). This disparity seen in the news media may result in fewer solved cases/ victim recovery for missing adults, minorities, men, or runaway youth; however this relationship is still being examined empirically. Although this disparity in news media attention has been empirically established, attempts to ameliorate this issue has proved difficult. Researchers suggest that long established news values, newsworthiness factors, and viewer engagement biases would require challenging in order to creating meaningful, wide scale reductions in disparity of news coverage (Jewkes and Linnemann 2018; Taylor et al. 2013). Intervening on these factors has proved quite difficult. In an attempt to increase exposure of missing persons cases, law enforcement has been moving to alternative means to increase awareness of missing persons cases that bypass the long held disparity in traditional news media coverage.

3 Alert Systems

Law enforcement agencies have taken steps to increase exposure of missing persons cases through the creation of two alert systems: American's Missing: Broadcast Emergency Response (AMBER) alert system and the Silver alert system. Both systems were created with the intention of relaying missing persons case information to the public quickly and efficiently (Zgoba 2004; Carr et al. 2009). Created in response to high profile missing children cases, the AMBER alert system works as a mechanism to relay information via television, radio, the internet, and text messages to the public in an attempt to increase awareness of certain types of missing children cases (Griffin et al. 2007; Zgoba 2004). The AMBER alert system is said to work as a crime control tool specifically to combat stereotypical stranger abductions, but is also utilized for other types of missing child cases (Griffin 2010; U.S. Department of Justice 2006). Specifically, AMBER alerts are relegated to cases involving individuals under the age of 18 who have been abducted and are at risk for danger or death (U.S. Department of Justice 2006). Specific information related to the characteristics of the victim, offender, and/or vehicle related to the crime is also an AMBER alert requirement (U.S. Department of Justice 2006). Similarly, the Silver alert system works to increase exposure/awareness of cases involving elderly individuals who are suspected of being missing due to cognitive impairment (Yamashita et al. 2013). Both of the missing persons alert systems were created with the intent of bypassing the reliance on traditional news media outlets in order to share information with the public and, hopefully, generate tips that result in the recovery of the victim (Carr et al. 2009; U.S. Department of Justice 2006; Yamashita et al. 2013; Zgoba 2004).

Unfortunately, there is a limited amount of research that examines the effectiveness of missing persons alert systems, as well as studies suggesting the ineffectiveness of the AMBER alert system. Preliminary evaluations of the AMBER alert system suggest that the alert system may work more to inflate perceptions of successful crime control, rather than function as a key crime control mechanism (Griffin 2010; Griffin and Miller 2008; Griffin et al. 2007). Specifically, Griffin examined over 200 resolved AMBER alert cases and found that the majority of cases that utilized the alert system were not stereotypical stranger abductions, rather familial abductions, suggesting that AMBER alerts may only have utility for certain types of cases (Griffin 2010; Griffin et al. 2007). Upon examination of the Silver alert system, preliminary results suggest an unequal application of the alert system, where location of the victim is associated with an increase in the likelihood of usage, as opposed to risk factors and case characteristics (Yamashita et al. 2013). There is also concern that the alert systems in place do not account for all of the missing persons cases that pose considerable threat to the victims (Griffin and Miller 2008), such as some missing adult and runaway youth cases. Taken together, research suggests that although there are current mechanism in place to relay missing persons information to the public, their utility and successfulness may be falling short to accommodate all real world missing persons cases.

4 Social Media as a Crime Control Alert System

Recently, law enforcement and government agencies, as well as non-profit organizations, have begun using social media as a means to share information quickly and with large audiences (Fletcher 2016). Social media use has become a common aspect of everyday human interaction for many US citizens (Edosomwan et al. 2011). In response to this near universal use of social media, agencies have begun to rely on social networking sites (SNS), such as Facebook and Twitter, to inform community members on a number of issues ranging from impending extreme weather and disaster relief to traffic and crime in the community (Bendler et al. 2014; Donaldson et al. 2013; Hattingh 2015; Milivojevic and McGovern 2014; Starbird 2011; Wang et al. 2012; Yin et al. 2012). Burgeoning research suggests that many law enforcement agencies are relying on social media to interact with community members, share information quickly, and increase citizen crime reporting (Beshears 2017; Brainard and Edlins 2015; Fletcher 2016; Hu et al. 2018; IACP 2018). Specifically, the International Association of Chiefs of Police (IACP) reports that as of 2018, 95.9% of law enforcement agencies use social media to interact with the public (IACP 2018). In addition, the majority of agencies who do not currently use social media are considering incorporating social media into their agency (IACP 2018). Furthermore, the Federal Bureau of Investigation (FBI) has also recommended that all law enforcement agencies establish a social media presence (Beshears 2017). This social media use can prove fruitful in aiding in police investigation, where the majority of agencies surveyed by the IACP report social media aided in solving crime in the community (Fletcher 2016).

Building from the research on social media engagement in the business and advertisement realm, research suggests that social media use by law enforcement agencies can work to increase exposure of crime information and level of community engagement (Colwyn 2014). Law enforcement agencies create content (posts) that encourage active user engagement (views, clicks, likes, reactions, and sharing of content) which increases user/community member exposure to the content at hand, thus garnering more attention for the crime related post (Colwyn 2014; Trainor 2012). As engagement increases, the amount of individuals who are capable of viewing the content also increases, thus resulting in additional exposure of the case (Colwyn 2014; Trainor 2012). Optimistically, this also results in an increase in the likelihood that potential witnesses solicit information and cases are resolved.

First, government and law enforcement agencies began to utilize social media as a means to relay information related to natural disasters and impending extreme weather alerts (Starbird 2011). Social media may work exceptionally well during disaster relief situations, where large groups of individuals can be kept informed and mobilized to act quickly (Gao et al. 2011; Yin et al. 2012). For example, SNS Twitter proved successful in relaying accurate information during a natural disaster, which reduced conflicting information, increased visibility, and allowed for the quick crowdsourcing of civilian aid (Starbird 2011). This finding was also replicated in a study on social media use during eight natural disasters in Australia, where Twitter worked as a valuable tool to alert the public and increase awareness and safety tactics (Yin et al. 2012).

Moving to social media as a tool for crime control, researchers have begun to assess the utility of this resource to relay accurate information to community members, in hopes to reduce crime and increase community awareness/safety (Bendler et al. 2014; Brainard and Edlins 2015; Hattingh 2015; Lieberman et al. 2013; Lock et al. 2013; Wang et al. 2012). SNS such as Twitter and Facebook are being utilized by law enforcement agencies in order to determine high crime areas and allow for community members to share information regarding crime problems within a community (Bendler et al. 2014; Brainard and Edlins 2015; Hattingh 2015; Lock et al. 2013; Wang et al. 2012). Moreover, Hattingh (2015) examined social media use in a South African community and found that social media use works to raise community awareness of particular crimes and increase self-reported community cohesiveness. Similarly, Kelly and Finlayson (2015) examined an online neighborhood watch program and found that social media aided in creation of community engagement and crime control at the neighborhood level. Furthermore, Linder and colleagues (2016) identified social media outreach via Facebook, Tumblr, and Instagram as a key mechanism in organizing group response and awareness of sexual assault on a college campus. Brainard and Edlins (2015) examined police initiated posts on several social media platforms and found that social media may work to increase positive police-community interactions and subsequently crime witness tips. Although limited in size, this growing body of research suggest that there is utility in law enforcement use of social media as a means to share crime information with the community.

5 Social Media as an Alert System for Missing Persons Cases

Taken together, this growing body of research suggests that social media is a valuable tool for crime control and the dissemination of important safety information to the community (Beshears 2017; Brainard and Edlins 2015; Fletcher 2016; Hu et al. 2018; IACP 2018; Lock et al. 2013; Starbird 2011; Wang et al. 2012; Yin et al. 2012). This movement into the social media realm may be especially relevant to missing persons cases, as social media has the ability to allow law enforcement agencies to relay information regarding a missing persons case quickly, to large groups of individuals, in a cost effective manner (Beshears 2017). Similar to traditional alert systems, social media works to relay accurate information to the public quickly and efficiently (Beshears 2017; Brainard and Edlins 2015; Starbird 2011; Yin et al. 2012). Considering the prevalence of social media use within the public, reliance on social media to relay crime related information appears to be a natural step in accommodating the lifestyle of the typical American citizen. In addition, social media may work as a means to overcome the limitations of traditional alert systems, in that social media posts are not restricted by the case criteria of AMBER and Silver alert systems. This lack of restrictions allows for a variety of missing persons cases, many of which were excluded from traditional alert systems, to receive exposure. Furthermore, social media may also work to ameliorate some of the disparity seen in traditional missing persons news media attention, as law enforcement agencies are not dependent on the beliefs and behaviors of large media outlets in order to relay information to the public. This may result in a variety of cases reaching audiences/community members that may not have historically received news media attention.

Social media as an alert system may also be a beneficial tool for dissemination of information related to runaway youth cases. Runaway youth comprise the largest amount of missing persons cases (NCIC 2017) and runaway cases can be particularly taxing on law enforcement, especially patrol officers (Chen et al. 2007). Considering this, a mechanism to aid law enforcement in the quick recovery of runaway youth is greatly needed. Social media has the potential to work as the only available alert system for locating runaway youth, in that their case information would not typically meet criteria for an AMBER alert (U.S. Department of Justice 2006) or meet criteria for newsworthiness of traditional news media (Taylor et al. 2013). To date, the only consistent system in place to relay information regarding runaway youth is the National Center of Missing and Exploited Children (NCMEC), which posts active runaway youth cases on their website (NCMEC 2019). Reliance on social media as a means to raise awareness of missing runaway youth within the community has the potential to rectify the limitations of previously established mechanisms for case exposure. Using social media, law enforcement agencies can generate exposure for runaway youth cases within the community and potentially increase the likelihood that these youth return home safely and quickly. Considering the high risk of exploitation and other negative life experiences during the runaway event (Chen et al. 2007; NCMEC 2015; Whitbeck and Hoyt 1999), additional tools to locate runaway youth is greatly needed. Reliance on social media as an awareness tool may be a beneficial, quick, and cost effective option for law enforcement agencies.

5.1 Missing Persons & Social Media Studies

Although still in its infancy, research suggests that social media usage by law enforcement may work as an effective and efficient alert system for missing persons case. In an examination of a law enforcement agency's social media use on the SNS Facebook, findings suggest that social media is an efficient tool to disseminate a variety of crime and community related information to the public (Jeanis et al. 2019). Jeanis et al. (2019) examined a sheriff's department's social media page for a 16 month period (July 2016–December 2017) in order to determine what types of posts are created and how they are engaged with by the public, with an in-depth focus on missing persons related content. Researchers examined levels of user engagement (minimum as clicks, moderate as likes/reactions, and maximum as shares) with missing persons, suspect wanted, and community/sheriff department related content (Jeanis et al. 2019). Several time of post related variables and type of post variables were also assessed in order to determine peak engagement times and types for the related content (Jeanis et al. 2019). Results of the law enforcement study suggest that, despite only accounting for 6% of the sample, Facebook posts related to missing persons cases are significantly more likely to receive social media engagement (clicks, likes, reactions, and shares) than all other types of posts (Jeanis et al. 2019). This finding supports the utility of social media to facilitate exposure of missing persons cases, as users were eager to actively and meaningfully engage with missing persons posts, thus increasing case exposure (Jeanis et al. 2019).

In addition, more in-depth analyses examined supplementary post factors that shape user engagement and related exposure for missing persons posts. For example, posts that include videos or pictures, as opposed to links, increase user engagement and may facilitate increased exposure of missing persons cases (Jeanis et al. 2019). This finding is consistent for a variety of post content, suggesting a powerful preference for posts that consist of photos or videos (Jeanis et al. 2019). This finding suggests a best practice guideline for law enforcement social media use, where social media users may have a strong preference for content created in a particular manner that has visual imagery and does not require the additional steps of clicking a link to additional content. Time variables were also considered for the study, and analyses suggest that posts created by the law enforcement agency were more likely experience user engagement when they were created after the typical work hours of 8am and 5pm (Jeanis et al. 2019). Considering this, it is suggested that law enforcement agencies may benefit from creating posts during particular time frames, specifically those that are outside of traditional US working hours. In summary, this study suggests that social media functions as a tool to increase exposure of a variety of missing persons cases and that users are likely to engage with missing persons posts.

Recently, an expanded analysis encompassing sheriff's department social media data, encompassing data from 2016–2018 confirms previous findings. Missing persons cases are significantly more likely to receive user engagement at all levels, compared to all other types of content, despite consisting of only 5% of the overall post content (Jeanis forthcoming). In addition, certain types of posts remain consistent predictors of user engagement, as posts that included videos are significantly more likely to receive engagement, which was particularly prominent for the sharing of content onto one's newsfeed (Jeanis forthcoming). These findings, consistent across three years of data, suggests that social media has the potential to be a valuable resource for law enforcement in the investigation of missing persons cases. In addition, these findings suggest the need to further examine what factors meaningfully shape engagement, in order to determine best practice guidelines.

6 Future Directions

Law enforcement agencies have begun to tap into the potential of social media to work as a crime control tool and alert system (Bendler et al. 2014; Brainard and Edlins 2015; Hattingh 2015; Lock et al. 2013; Wang et al. 2012); however, this field of study is still in its infancy. Future research is needed in order to better understand the utility of social media for collecting evidence, soliciting witness information, and increasing awareness of missing persons cases. Time and exposure are key factors in missing persons cases (Hickey 2016); therefore, the empirical examination of how social media can work to reduce the length missing and increase awareness of cases is greatly needed. Although considerable evidence strongly suggests the utility of social media (Brainard and Edlins 2015; Kelly and Finlayson 2015; Jeanis et al. 2019), and most law enforcement agencies have already begun to use social media (IACP 2018), empirical investigation of the use, as well as best practice guidelines, is greatly underdeveloped.

6.1 *Broad Research Development*

Law enforcement agencies speak to the use of social media (as seen in the IACP 2018 survey); however, studies assessing the specific ways in which agencies commonly employ social media as a tool for crime control is greatly needed. Exploratory research that examines the means and motivations for using social media in the realm of law enforcement is limited at best, with Hu et al. (2018) being one of the few studies that assess the variety of ways that law enforcement agencies use social media. Similarly, it is unknown what types of users engage with law enforcement in the social media realm, or their motivations. Typically, social media artifacts are examined (engagement as likes, clicks, shares, comments, views) but direct interaction with and experiences of users is limited at best (see Hattingh 2015). Future researchers should endeavor to survey those who participate in social media interaction with law enforcement in order to identify additional aspects of this relationship.

Moreover, the relationship between social media and evidence collection/case closure needs additional examination. Some studies suggest that social media can work to increase the likelihood of crime information being provided to law enforcement (Brainard and Edlins 2015); however, quantifying that this has yet to occur. An additional aspect that greatly needs development is the identification of best practices for law enforcement social media use. Although agencies may have logistical guides on how to create content on Facebook or Twitter, specific ways to best utilize social media as a crime control tool (best practice guidelines) are completely lacking in this field of study. This is a considerable concern because agencies are utilizing a resource of great potential; however, they may not have the expertise or knowledge to inform usage and increase its utility as a crime control tool and alert system. Factors related to traditional media attention may also serve to aid law enforcement in understanding how to increase exposure of cases. For example, traditional newsworthiness principles may also be applied to social media settings, and this would be knowledge that the average law enforcement agent who not inherently possess, thus needing training. Similar to other technological training programs, a social media training guide and related training program could greatly benefit law enforcement agencies.

6.2 Missing Persons Research Development

Moving to specific application of social media studies to missing persons investigations, additional research in this area of interest is greatly needed. Considering the unique relationship between the media and missing persons cases, as well as the growing body of literature that suggests social media's utility (see Jeanis et al. 2019), additional research that examines this relationship has the potential to benefit the field of study and larger societal forces. Specifically, the examination of a large geographically diverse sample of law enforcement agencies' use of social media for missing persons investigations would aid in the understanding of how various agencies are using social media as an alert system. In addition, the role of social media in increasing solvability of cases and recovery of victims is greatly needed. For example, comparing cases that receive social media attention to those missing persons cases that do not would serve to better identify the role of social media as a mechanism to increase likelihood of case closure and victim recovery. Furthermore, quantifying and qualitatively examining witness tips provided through social media (or as a result of social media awareness) as they relates to missing persons cases would also aid in understanding how social media works as an alert system/crime control tool.

Law enforcement use of social media to raise awareness of missing persons cases may be an particularly beneficial tool for cases that are unlikely to receive traditional media attention, or meet criteria for traditional alert systems. For example, African American and ingenious women, runaway youth, and adult men are less likely to receive media attention (Gilchrist 2010; Jeanis and Powers 2017; Min and Feaster 2010; Taylor et al. 2013) and may not meet criteria for the AMBER alert system; however, a social media based alert system that can link to local communities and larger organizations may work to rectify this issue. Research to test the applicability for these types of cases is greatly needed, as social media may hold untapped potential for law enforcement. Future studies should aim to examine if social media works to garner exposure for these types of cases, as well as the amount and type of coverage as it relates to case closure/victim recovery. Considering the current revelation that indigenous women in the US are being obscured in data collection on a national scale, as well as concern regarding the integrity of missing persons investigations (Lucchesi and Echo-Hawk 2018), resources and tools to combat this issue and raise exposure of these cases are needed. Social media may work as a cost effective mechanism to increase community awareness for law enforcement agencies that have limited resources. This may be especially relevant for tribal nations, who historically have reduced access to traditional media outlets and crime control tools/funds (Lucchesi and Echo-Hawk 2018).

In conclusion, considering the prominence of social media, as well as its ability to reach large groups of individuals quickly, social media can be a beneficial tool for law enforcement agencies. A growing body of research suggests the utility of social media as a crime control tool and alert system (Bendler et al. 2014; Brainard and Edlins 2015; Hattingh 2015; Jeanis et al. 2019; Lock et al. 2013; Starbird 2011;

Wang et al. 2012; Yin et al. 2012). Specifically, social media is an optimal tool for increasing awareness of missing persons/runaway youth cases, and increased usage as well as empirical assessment is recommended. This growing field of research may be on the cusp of transforming the ways in which law enforcement agencies interact with the community, solicit witness tips, and raise awareness of missing persons cases. Research on the role of social media in criminal investigations and outreach is growing, as law enforcement continues to adapt to the changing social structure of society today.

References

Bendler, J., Brandt, T., Wagner, S., & Neumann, D. (2014). Investigating crime-to-Twitter relationships in urban environments-facilitating a virtual neighborhood watch. In *Twenty Second European conference on information systems*. Tel Aviv, Israel.

Beshears, M. L. (2017). Effectiveness of police social media use. *American Journal of Criminal Justice, 42*(3), 489–501.

Biehal, N., Mitchell, F., & Wade, J. (2003). *Lost from view: Missing persons in the UK*. Bristol: The Policy Press.

Brainard, L., & Edlins, M. (2015). Top 10 US municipal police departments and their social media usage. *The American Review of Public Administration, 45*(6), 728–745.

Carr, D., Muschert, G. W., Kinney, J., Robbins, E., Petonito, G., Manning, L., & Brown, J. S. (2009). Silver alerts and the problem of missing adults with dementia. *The Gerontologist, 50*(2), 149–157.

Chen, X., Thrane, L., Whitbeck, L. B., Johnson, K. D., & Hoyt, D. R. (2007). Onset of conduct disorder, use of delinquent subsistence strategies, and street victimization among homeless and runaway adolescents in the Midwest. *Journal of Interpersonal Violence, 22*(9), 1156–1183.

Colwyn, S. (2014). *Research summary: Social media marketing effectiveness in 2014*. Smart Insights.com. Retrieved from: https://www.smartinsights.com/social-media-marketing/social-media-governance/social-media-marketing-effectiveness-2014/

Donaldson, O., Duggan, E., & Maitland, N. (2013). Using a design science approach to create and evaluate a social media crime reporting tool in a developing country: Case Jamaica. In *Americas conference on information systems*. Chicago, IL.

Edosomwan, S., Prakasan, S. K., Kouame, D., Watson, J., & Seymour, T. (2011). The history of social media and its impact on business. *Journal of Applied Management and Entrepreneurship, 16*(3), 79.

Fletcher, B. O. (2016). *Social media: An indispensable tool for law enforcement* (Unpublished white paper). Leander: Leadership Command College.

Fyfe, N. R., Stevenson, O., & Woolnough, P. (2015). Missing persons: The processes and Challenges of police investigation. *Policing and Society, 25*(4), 409–425.

Gao, H., Barbier, G., & Goolsby, R. (2011). Harnessing the crowdsourcing power of social media for disaster relief. *IEEE Intelligent Systems, 26*(3), 10–14.

Gilchrist, K. (2010). 'Newsworthy' victims? Exploring differences in Canadian local press coverage of missing/murdered Aboriginal and White women. *Feminist Media Studies, 10*(4), 373–390.

Greene, K., & Pakes, F. (2014). The cost of missing persons investigations: Implications for current debates. *Policing: A Journal of Policy and Practice, 8*(1), 27–34.

Griffin, T. (2010). An empirical examination of AMBER Alert 'successes'. *Journal of Criminal Justice, 38*, 1053–1062.

Griffin, T., & Miller, M. (2008). Child abduction, AMBER alert, and crime control theater. *Criminal Justice Review, 33*(2), 159–176.

Griffin, T., Miller, M., Hoppe, J., Rebideaux, A., & Hammack, R. (2007). A preliminary examination of AMBER Alert's effects. *Criminal Justice Policy Review, 18*(4), 378–394.

Hattingh, M. (2015). The use of Facebook by a community policing forum to combat crime. In *Proceedings of the 2015 annual research conference on South African Institute of Computer Scientists and Information Technologists*. ACM, South Africa.

Hickey, E. W. (2016). *Serial murderers and their victims*. Cengage.

Hu, X., Rodgers, K., & Lovrich, N. (2018). "We are more than crime fighters": Social media images of police departments. *Police Quarterly, 21*(3), 1–29.

Jeanis, M. (forthcoming). *Social media use by law enforcement: Recommendations for best practices*.

Jeanis, M. N., & Powers, R. A. (2017). Newsworthiness of missing persons cases: An analysis of selection bias, disparities in coverage, and the narrative framework of news reports. *Deviant Behavior, 38*(6), 668–683.

Jeanis, M. N., Muniz, C. N., & Molbert, C. L. (2019). Law enforcement and social media usage: An analysis of engagement. *Policing: A Journal of Policy and Practice*. https://doi.org/10.1093/police/paz026.

Jewkes, Y., & Linnemann, T. (2018). *Media and crime in the US*. Thousand Oaks: Sage Publications.

Kelly, A., & Finlayson, A. (2015). Can Facebook save neighbourhood watch? *The Police Journal, 88*(1), 65–77.

Lieberman, J. D., Koetzle, D., & Sakiyama, M. (2013). Police departments' use of Facebook: Patterns and policy issues. *Police Quarterly, 16*(4), 438–462.

Linder, C., Myers, J. S., Riggle, C., & Lacy, M. (2016). From margins to mainstream: Social media as a tool for campus sexual violence activism. *Journal of Diversity in Higher Education, 9*(3), 231.

Lock, R., Cooke, L., & Jackson, T. (2013). Online social networking, Order and disorder. *The Electronic Journal of E-Government, 11*(1), 229–240.

Lucchesi, A., & Echo-Hawk, A. (2018). *Missing and murdered indigenous women and girls: A snapshot of data from 71 urban cities in the United States*. Seattle: Urban Indian Health Institute.

Milivojevic, S., & McGovern, A. (2014). The death of Jill Meagher: Crime and punishment on social media. *International Journal for Crime, Justice and Social Democracy, 3*(3), 22–39.

Min, S., & Feaster, J. (2010). Missing children in national news coverage: Racial and gender representations of missing children cases. *Communication Research Reports, 27*(3), 207–216.

Morewitz, S., & Colls, C. (Eds.). (2016). *Handbook of missing persons*. Cham: Springer International Publishing.

Moscowitz, L., & Duvall, S. (2011). "Every parent's worst nightmare" Myths of child abductions in US news. *Journal of Children and Media, 5*(02), 147–163.

National Center of Missing and Exploited Children (NCMEC). (2015). *Missing children, state care, and child sex trafficking: Engaging the judiciary in building a collaborative response*. Alexandria: National Center for Missing & Exploited Children and National Council of Juvenile and Family Court Judges.

National Center of Missing and Exploited Children (NCMEC). (2019). *Search for missing children*. Retrieved from: http://www.missingkids.org/gethelpnow/search

National Crime Information Center (NCIC) Missing Person and Unidentified Person Statistics for 2017. (2017). *Federal Bureau of Investigation*. Retrieved from: https://www.fbi.gov/file-repository/2017-ncic-missing-person-and-unidentified-person-statistics.pdf/view

Newiss, G. (2005). A study of the characteristics of outstanding missing persons: Implications for the development of police risk assessment. *Policing and Society, 15*(2), 212–225.

Simmons, C., & Woods, J. (2015). The overrepresentation of White missing children in national television news. *Communication Research Reports, 32*(3), 239–245.

Starbird, K. (2011, May). Digital volunteerism during disaster: Crowdsourcing information processing. In *Conference on human factors in computing systems* (pp. 7–12).
Taylor, J., Boisvert, D., Sims, B., & Garver, G. (2013). An examination of gender and age in print media accounts of child abductions. *Criminal Justice Studies, 26*(2), 151–167.
The Commission on Peace Officer Standards and Training. (2011). *Missing persons investigations & guidelines curriculum.* California: The California Commission on Peace Officer Standards and Training.
The International Association of Chiefs of Police (IACP). (2018). *Social media.* IACP. https://www.theiacp.org/resources/policy-center-resource/social-media.
Trainor, K. J. (2012). Relating social media technologies to performance: A capabilities-based perspective. *Journal of Personal Selling & Sales Management, 32*(3), 317–331.
U.S. Department of Justice. (2006). *AMBER Alert: America's missing: Broadcast emergency response*. Amber Alert. https://amberalert.ojp.gov/.
Wang, X., Gerber, M., & Brown, D. (2012). Automatic crime prediction using events extracted from twitter posts. In *International Conference on Social Computing, Behavioral-Cultural Modeling, and Prediction*. Berlin, DE.
Whitbeck & Hoyt. (1999). *Nowhere to grow*. Hawthorne: Walter de Gruyter, Inc.
Windle, M. (1988). Substance use and abuse among adolescent runaways: A four-year follow-up study. *Journal of Youth and Adolescence, 18*(4), 331–344.
Yamashita, T., Carr, D. C., & Brown, J. S. (2013). Analyzing state-based Silver Alert programs: The case of North Carolina. *North Carolina Medical Journal, 74*(2), 111–117.
Yin, J., Lampert, A., Cameron, M., Robinson, B., & Power, R. (2012). Using social media to enhance emergency situation awareness. *IEEE Intelligent Systems, 27*(6), 52–59.
Zgoba, K. (2004). The Amber alert: The appropriate solution to preventing child abduction? *The Journal of Psychiatry & Law, 32*(1), 71–88.

Police Engagement in Multidisciplinary Team Approaches to Commercial Sexual Exploitation of Children

Amy Farrell, Candence Wills, and Carlande Nicolas

Abstract Over the past two decades law enforcement has been given new mandates to identify and respond to the commercial sexual exploitation of children (CSEC). While CSEC has been identified as an important community safety issue and the negative impacts on young people are well documented, it has proven to be difficult for the police to identify young people who are at risk of victimization and challenging to develop the trust necessary to gather information needed to support investigations, arrest perpetrators, and provide safety to child victims. In response to these challenges, law enforcement is increasingly partnering with child welfare and service provider stakeholders to improve responses to CSEC in local communities. This chapter describes the problem of CSEC and outlines key aspects of a multidisciplinary team approach to the problem. Using qualitative interview data from participants in six multidisciplinary teams in Massachusetts dedicated to responding to CSEC, we discuss key benefits and challenges to a multidisciplinary team approach to CSEC.

Keywords Commercial sexual exploitation of children · Child trafficking · Sex trafficking · Multidisciplinary teams · Police · Victims

1 Introduction

The commercial sexual exploitation of children (CSEC), sometimes referred to as child sex trafficking has become an increasingly important concern in local communities. Federal law, including the Victims of Trafficking and Violence Protection Act of 2000 (TVPA 2000) and the Justice for Victims of Trafficking Act (JVTA

A. Farrell (✉) · C. Wills · C. Nicolas
School of Criminology and Criminal Justice, Northeastern University, Boston, MA, USA
e-mail: am.farrell@northeastern.edu

B. Fox et al. (eds.), *Science Informed Policing*, Advanced Sciences and Technologies for Security Applications,
https://doi.org/10.1007/978-3-030-41287-6_10

2015) identifies minors who are engaged in commercial sex as victims of sex trafficking regardless of coercion, fraud, or force (TVPA 2000, P.L. 106–386) and expands the definition of sex trafficking to include sexual exploitation and other sexual abuse types, pornography, live or online sex shows, or sex tourism (JVTA 2015, amending 18 U.S.C. 1591[a][1]). All 50 states have enacted legislation prohibiting sex trafficking, with a majority of states including specific provisions responding to the commercial sexual exploitation of minors (Shared Hope 2016). The terms sex trafficking and commercial sexual exploitation are sometimes used interchangeably, but throughout this chapter we use the broader term of commercial sexual exploitation.

There are no accurate estimates of the prevalence of commercial sexual exploitation in the U.S., but the number of young people at risk for exploitation is significant. Approximately 3.5% of a nationally-representative sample of middle and high school youth reported they had exchanged sex for money or drugs in their lifetime (Edwards et al. 2006). Young people who experince CSEC experience significant negative health outcomes as a result of their exploitation including posttraumatic stress disorder (PTSD), anxiety and depression, substance abuse, anger, self-harm or suicidal ideation, sexualized behavior, cognitive impairment or memory problems (Cole et al. 2016; Landers et al. 2017; Le et al. 2018; Varma et al. 2015).

Law enforcement has assumed greater responsibility in responding to commercial sexual exploitation over the past two decades. In 1998, the Office of Juvenile Justice and Delinquency Prevention (OJJDP) created the Internet Crimes Against Children (ICAC) Task Force Program under the 1998 Justice Appropriations Act (OJJDP 1998; Public Law 105–119) to combat use of the internet and other technologies perpetuating the victimization of children by developing responses to online sexual predators, obscene and pornographic images of minors, and youth sexual exploitation (USDJ 2010; OJJDP n.d.). The ICAC program encompasses a "national network of 61 coordinated task forces representing over 4500 federal, state, and local law enforcement and prosecutorial agencies to assist in investigation, forensic examinations, and prosecution of internet crimes against youth" (ICAC 2017). Congress and President Bush in 2008 signed the Providing Resources, Officers, and Technology to Eradicate Cyber Threats to (PROTECT) Our Children Act of 2008 which authorized the ICAC Task Force Program (USDJ 2008; Public Law 110–401). Based on the success of the ICAC model, the federal government utilized a multidisciplinary team approach to support local responses to sex trafficking and CSEC. Since the passage of the TVPA in 2000, the federal government has devoted significant resources to supporting the training of local, county and state law enforcement officials about sex trafficking with an emphasis on the commercial sexual exploitation of children. Local task forces were funded to enhance law enforcement collaboration with service providers and a range of local stakeholders to promote a holistic response to human trafficking victims, including minor victims of commercial sexual exploitation (Bureau of Justice Assistance n.d.).

To bolster local efforts in identifying and responding to child victims, in 2014, the federal government took steps to clarify the responsibilities of local authorities in the identification of and response to commercial sexual exploitation. The

Preventing Sex Trafficking and Strengthening Families Act (2014, Title IV-E, Section 101) mandates that state child welfare agencies "develop policies and procedures to identify, document, screen, and determine appropriate services for children who are victims of, or at risk of, sex trafficking or a severe form of trafficking in persons." Under this mandate, law enforcement must work collaboratively with child welfare agencies and other providers when CSEC victims are identified and in providing responses that ensure the continued safety of these victims. A mechanism for ensuring adequate response to CSEC victims is the collaboration of law enforcement with child welfare and other providers through a multidisciplinary team response.

Despite the passage of law and the allocation of resources, law enforcement has struggled in many communities to identify victims and establish effective responses to meet victim needs (Farrell et al. 2019). Although there are no statistics about the number of CSEC cases that have been identified by the police, the Uniform Crime Reporting program does track the broader category of sex trafficking offenses known to police, which includes both adult and minor victimization. In the most recently released statistics, local law enforcement agencies reporting into the UCR identified just under 1000 (994) sex trafficking offenses in 2017. The number of offenses identified by the police is dwarfed by estimates of the problem of sex trafficking of adults and minors suggesting hundreds of thousands of victims in the Americas (ILO 2017).

The challenges facing law enforcement in successfully identifying CSEC victims are well documented (Farrell et al. 2014, 2015, 2019; Farrell and Pfeffer 2014; Newton et al. 2008). Research suggests that police responses are complicated when victims of commercial sexual exploitation are reluctant to disclose information about victimizations out of fear of harm by the offender or facing criminal charges for activities related to their victimization (O'Brien et al. 2017; Reid et al. 2017; Aron et al. 2006; Love et al. 2018). Additionally, young people may not disclose CSEC victimization due to their emotional attachments to the person exploiting them, fear of being judged by others, and self-blame (Andretta et al. 2016). In addition to challenges related to the crime and victim reluctance to disclose to authorities, law enforcement agencies lack critical training about CSEC or sex trafficking (Bjelland 2017; Farrell et al. 2010; Grubb and Bennett 2012; Irwin 2017) and need assistance from collaborative partners to enhance both identification and response.

2 Background on Multidisciplinary Teams

The multidisciplinary team (MDT) model is nationally recognized as an effective response to challenging and complex social welfare issues (Epstein and Edelman 2014; Cole 2018). Within social services and the juvenile justice system, MDTs are becoming more common in particular with addressing investigations in cases of child abuse, maltreatment and neglect, youth trauma, and adolescent behavioral intervention (Jacobson 2001; Cheung and Brandes 2011; Brink et al. 2015; Bracewell 2018). Frequently, these complex cases involve the intersection of multiple public systems which necessitates cross-agency collaboration. Members of MDTs addressing child abuse may vary but largely comprise of law enforcement, prosecution, medical professionals, mental health providers, and personnel in family advocacy and child protective services (Lalayants and Epstein 2005; Bracewell 2018). Each MDT creates a community of diverse professionals who bring unique contributions to collectively provide an individualized plan of action to best address youth victimization.

Representing various fields and agencies within the public systems, members of multidisciplinary teams work in tandem to tackle a case issue with a coordinated response. A distinct feature of the Child Advocacy Center (CAC) model when introduced in the 1980s, the CAC model has shown to result in positive outcomes, including enhancing communication across service agencies (Newman et al. 2005; Fryer et al. 1988; Skaff 1988; Tjaden and Anhalt 1994) and increasing coordination and collaboration among stakeholders (Bai et al. 2009; Cross et al. 2005; Hurlburt et al. 2004; Kolbo and Strong 1997; Cross et al. 2007; Faller and Palusci 2007). Overall, MDTs have improved outcomes by streamlining processes in the criminal justice system, providing more comprehensive child protections and services, and increasing satisfactions among all team members involved.

Coordination between law enforcement, prosecutors, and child advocacy centers within MDTs increases the rate and quality of successful prosecutions and arrests of child abuse cases (Miller and Rubin 2009; Dinsmore 1992; Bradford 2005; Joa and Edelson 2004; Turner 1997; Feng et al. 2010). With greater coordination and information sharing, teams improve identification and investigation of commercial sexual exploitation cases, leading to a rise in solvable cases (Hochstadt and Harwicke 1985; Kolbo and Strong 1997). As well, multidisciplinary teams encourage greater law enforcement participation in cases (Cross et al. 2007; Smith et al. 2006) and further lead to greater substantiation for child abuse cases (Wolfteich and Loggins 2007).

Employing MDTs boosts the accuracy of risk assessments and predictions (Goldstein and Griffin 1993; Pence and Wilson 1994). MDTs expand the number of case reviews by recognizing 'red flags' or indicators of abuse and victimization (Cross et al. 2007). Those utilizing services claim that services are more cohesive and available (Onyskiw et al. 1999). In one study, after working in a team setting law enforcement argued in favor of using MDTs to assist with referrals to child advocacy agencies, counseling and mental health support, and medical services (Newman et al. 2005).

From the perspective of caregivers and children who benefit from public services, they are more inclined to be satisfied following the CAC model (Jones et al. 2007). Professionals reportedly convey that the MDTs enhance client relations and overall team confidence in approaching a case (Fryer et al. 1988; Bross et al. 2000). Team members and associates also report having significantly greater satisfaction with MDT consultation on cases (Lalayants et al. 2011), while agency staff members are more satisfied with the new level of child protection (Goldbeck et al. 2007).

2.1 MDT Response to Commercial Sexual Exploitation of Children

Reports show that the MDT model is useful in responding to cases concerning commercial and sexual exploitation of children (CSEC). Too often, youth who are bought and sold for sexual services may be treated as offenders and arrested for prostitution (Epstein and Edelman 2014). MDTs that address CSEC shift the image of youth survivors away from offenders deserving of punishment towards victims in need of services. Teams can then coordinate care and treatment through public service providers for exploited youth, rather than labeling the youth as "prostitutes" and processing the victims through the juvenile or criminal justice systems. Further, having collaboration between team members and establishing a singular forensic interview is shown to more effectively support criminal investigations (Herbert and Bromfield 2017).

Similar to child abuse cases, youth involved in commercial sexual exploitation (CSE) often have prior contact with multiple public service systems. CSE youth may have open Department of Children and Families (DCF) cases for status offenses such as running away from home, truancy, or a familial history of domestic violence. Moreover, their victimization may extend across jurisdictions. Several agencies that work with vulnerable populations within the MDT model improve identification of these youth already established in their systems. Child victims of CSE may be more easily recognized in a system that is aware of potential CSE indicators and works together in identifying at risk youth in public systems (Epstein and Edelman 2014). Training teams on CSEC victim awareness and identifying 'red flags' has also shown to improve identification of abused children and hidden CSE youth victims (Lalayants 2013).

Challenges naturally arise when building teams from multiple public agencies in criminal justice and social service sectors. Altering workplace culture from independent agency responses to convening as a team before preemptively taking action can be met with resistance by members of the team (Johnson 2013).When first implementing a collaborative team response, members can be wary of drastic changes, such as moving from paper to electronic systems for tracking or filing processes. One study exhibits team members having varying levels of participation, highlighting clear inequalities in levels of participation with members from the prosecutor's office dominating team meetings (Bell 2001).

To avoid potential pitfalls when building MDTs, participants on the specialized team or task force create and establish a team protocol. Successful protocols are unique to the team and outline the specific roles for each member, procedures for convening, and processes of negotiating an action plan for each case (Moran et al. 2007). By incorporating communication opportunities for team members to interact in-person in both formal and informal settings, MDTs reduce barriers in participation and build trust within teams (Moran et al. 2007; Johnson 2013). Creating a balanced and enthusiastic team takes time and encouragement. For individuals to benefit from a collaborative response, it is necessary for individuals to engage with a supportive environment, particularly when team members experience "burn out" from being overburdened with stressful and straining caseloads (Lalayants 2013). For any sustainable model to improve responses to CSE of children, the programs and procedures implemented need to be financially supported. As more victims are identified, caseloads increase in the public system, creating more work for team members (Moran et al. 2007). Through an increase in funding allocated to the development of MDTs, teams can expand in capacity to alleviate the burgeoning caseloads and resource strain. Governments that provide funding provisions for MDTs send a signal to agencies and the public that the issue is not just a symbolic gesture, but a public priority and generate pressure to implement effective programs.

2.2 Police Engagement in MDTs

Participation in multidisciplinary teams to address crime and public order problems are not new to police. Such models have been utilized to address various crimes including drug trafficking (High Intensity Drug Trafficking Areas), guns (Project Safe Neighborhood), and gun-related violence (Safe Streets Violence Crime Initiative, Strategic Approaches to Community Safety Initiative). Multidisciplinary teams are by their very nature temporary organizations (Scott 1998). They tend to be time bounded, task specific and involve multiple agencies that often have different, and in some cases competing goals. In their analysis of multi-agency drug teams in Illinois, Olson et al. (2002) note that while there is considerable diversity in law enforcement engagement in multidisciplinary teams (sometimes referred to as task forces), team composition, and team function and goals, these teams share many common features. These include collaboration, the expanded exercise of discretion (particularly the use of federal statutes to prosecute what had been regarded as "state" crimes), and increased use of problem-solving techniques.

Within this context, the creation and maintenance of multidisciplinary teams that are by their very nature temporary can be difficult. Doing so requires coordination of activities, services, and communication across various levels of jurisdiction, for law enforcement this includes local, state, and federal partners in the team), across participant role and function (e.g. a prosecutor and a legal advocate), and across groups with various funding constraints. One of the major challenges of such multidisciplinary team approaches is that the participants see the group process as time limited, and the individual participants' reward structures are linked to their home agency or organization, not the multidisciplinary team. For example, police participating in an MDT may work toward the goal of securing evidence and making an arrest while these outcomes may be unimportant to other team members who are focused on providing stability and resources to victims. As a result, building trust and establishing a set of cultural norms within the multidisciplinary team, and developing group goals that everyone agrees to work towards can be challenging (Decker et al. 2005; Maguire and Katz 2002).

Although multidisciplinary team structure (e.g. leadership, regular meetings, established group process or protocols) is necessary, it is not sufficient alone to overcome the challenge of a temporary group without attention being paid to the cultural issues that each team member brings to the group. A critical element to overcoming this challenge is both formal and informal communication among group members and among home agency leadership. The individual communication techniques, protocols, styles and habits that MDT members bring to the temporary group need to align with any new communication structure established by the team. Research shows that increased communication and information sharing, enhanced opportunities for teamwork, and increasing the social capital among members through training are strategies to overcome the challenges of temporary organizations (Smith 2011; Twomey et al. 2010; Ruhstaller et al. 2006). Decker et al. (2005) utilized the case of the multidisciplinary team tasked with providing security services at the Salt Lake Olympics as an example of a group that overcame difficulties inherent in forming and implementing a team approach to a problem. As they note, these tasks are difficult enough in settings that exclusively involve law enforcement. MDTs designed to respond to commercial and sexual exploitation of children add considerable layers of organizational complexity to this already difficult task as they attempt to integrate multiple functions among team members from a variety of disciplines including child welfare, victim services, schools, and the medical community in addition to law enforcement, such as police and prosecution. In many cases agencies participating in the teams may have worked on opposite sides of issues in the past or may define the problem facing young people in different ways.

2.3 Case Study: A Multidisciplinary Team Approach to Child Trafficking in Massachusetts

To more fully understand the multidisciplinary team response to child trafficking, we examine the case of a multidisciplinary team response model that has been developed to identify and respond to CSEC in every county in Massachusetts. First, we describe the model generally to understand its components and implementation. Second, utilizing interviews from practitioners participating in multidisciplinary teams in six counties, we examine how having an MDT response model changed the ways practitioners in each county approached the problem of child trafficking. We also explore the challenges of implementing and operating the multidisciplinary team response model. Six counties were chosen for study because they were the first counties to adopt the MDT model outside of a county in the Boston-metro area, which had been using the MDT model for over a decade.

In each study site, the research team conducted in-depth, semi-structured interviews with a variety of practitioners who were involved in the MDT implementation and operation. These included Children's Advocacy Center directors, child trafficking coordinators, county and federal prosecutors, managers and caseworkers from the

Department of Children and Families, detectives and other law enforcement officials, and residential placement program directors or managers. In total, 36 interviews were conducted with MDT partners across the six counties between 2016 and 2018. Interview participants were selected using a purposeful sampling approach based on experience with child trafficking responses and service provision; informed consent was obtained from all individual participants included in the study, all respondents were ensured confidentiality, and participation in the interviews was voluntary.

All interview audio files were transcribed, and interview text was uploaded into QSR-NVivo (QSR International Pty Ltd., Victoria, Australia), a qualitative data analysis software package for coding and analysis. Thematic codes were developed to represent common themes derived from interviewers. Preliminary themes were developed based on literature reviews about multidisciplinary team responses. New themes were also added that emerged from the interview coding. These themes were analyzed within and across participants and study sites to identify any differences in the patterns of themes. We developed analytic memos to explore each emerging theme in depth.

3 The MDT Approach to Commercial Sexual Exploitation of Children in Massachusetts

The problem of commercial sexual exploitation of children was formally recognized by local leaders in Massachusetts in 2001 following the tragic death of a young women who had been placed in a group home under the care of the state child welfare agency after being arrested for prostitution. This tragedy brought stakeholders in the state together to improve the response to commercially exploited youth who had historically fallen through the cracks of most protective services. In the Boston-metro area engagement around CSEC led to the creation a county-wide model of multidisciplinary response to commercial sexual exploitation of children developed by the Children's Advocacy Center of Suffolk County. The multidisciplinary team organized by the CAC had buy-in from the leadership of local stakeholders. Central to the MDT approach was the coordination of responses to young people who were at risk for CSEC by a coordinator housed at the CAC who convened team meetings with MDT members involved in the lives of young people at risk. The teams developed safety plans, agreed on approaches to improve residential stability and coordinated forensic interviews or communication with the victimized youth.

Ten years later the Massachusetts legislature passed comprehensive anti-human trafficking legislation that strengthened protections for victims of sex trafficking and increased the punishment for offenders. It also specified that every child identified as a victim of CSEC would be provided an MDT response to meet their needs and increase their safety. This provision specifies:

> For 51A reports [state child abuse allegation reports] specifically involving a sexually exploited child or a child who is otherwise a human trafficking victim, the multi-disciplinary service team may consist of a team of professionals trained or otherwise experienced and

qualified to assess the needs of sexually exploited children or children who are otherwise human trafficking victims including, but not limited to, a police officer, as defined by section 1 of chapter 90C, or other person designated by a police chief, as defined in said section 1 of said chapter 90C, an employee of the department of children and families, a representative of the appropriate district attorney, a social service provider, a medical professional or a mental health professional. The team shall review and monitor the service plan developed by the department under subsection (g) of section 51B. The team shall evaluate the effectiveness of the service plan in protecting the child from further abuse or neglect. The team shall make recommendations regarding amendments to the service plan, the advisability of prosecuting members of the family, and the possibility of utilizing diversionary alternatives. If the team finds that services required under such plan are not provided to the family, the case shall be referred to the commissioner. (Massachusetts General Laws, Part I, Title XVII, Chapter 119, Section 51D)

Partnerships among numerous state agencies supported by a grant from the U.S. Department of Health and Human Services supported the development and implementation of multidisciplinary teams to respond to child trafficking in each county in the Commonwealth by 2018. MDT teams were coordinated by the Children's Advocacy Center in each county. Although team members varied by the needs and capacities in each county, core team members included representatives from the CAC, District Attorney's Office, local, state and federal law enforcement, department of child welfare, residential placement programs, probation, medical providers and local youth serving agencies. Technical assistance efforts supported by the grant included training of front-line child welfare and MDT partners to increase awareness and understanding of what the issue of CSEC looks like and how to respond to a youth victim. Additionally, MDT partners were convened for multi-day trainings to establish the protocol and goals of the MDT, increase awareness and understanding of both CSEC and various MDT partner roles in addressing the problem, and establish guidelines for information sharing. Finally, technical assistance was provided after each county-based MDT was convened to address concerns about youth safety and placement and to enhance investigation and prosecution of human trafficking cases. To support these county-based MDT efforts, CSEC case coordinators were funded in each county to manage the intake of child exploitation cases, coordinate communication among MDT members, and oversee the work of each of the county-based teams.

4 How the MDT Model Changed Approaches

4.1 Identification and Awareness

Implementing multidisciplinary teams in Massachusetts raised awareness of the issue of child exploitation in local communities and provided practitioners the tools to identify possible CSEC cases. As a family advocate in one study site states when referring to how the MDT model changed their knowledge of CSEC in their communities, "I think it's really to increase identification of these girls, increase awareness of the problem in the area to ensure that we're having—coming at it from a victim centered perspective" (Family Advocacy Center Program Manager, County 1). Educating both the wider community of practitioners and the specific members of the MDT about CSEC and sharing information regarding potential CSEC cases, increased the identification of victims who may have previously gone unrecognized by law enforcement. As one county CAC director noted, "I think the CSEC initiative is definitely offering law enforcement a tool, a new way of looking at these cases and I would hope that with the training and developing their skill set, we're looking at local girls and it's going to open their eyes when it comes to…you know it's not going to be just another runaway" (CAC Director, County 3). Previous to building an MDT, local agencies may have encountered young people who were victims of exploitation without recognizing it and without the tools to respond appropriately. As a CSEC case coordinator noted, "I would say the biggest change is just in the level of knowledge and coordination. So, I think that identification is key because we had some [what we now see as] very clear CSEC cases prior to the implementation of the grant" (CSEC Coordinator, County 2).

Training MDT members about CSEC and continuing to educate and work through identification challenges through the process of case coordination has improved members' ability to recognize CSEC as a local problem appearing in their own caseloads. While discussing the process of training local providers, particularly those from child welfare, about child sexual exploitation, one MDT member from law enforcement highlighted the importance of awareness and CSEC recognition as a concurrent issue in caseloads:

> The point I make to them [police] right out of the gate was well you guys have to understand that this isn't a different problem, it's the same demographic. My victims are your victims. They're the same people. You just need to have this awareness level to wear like this additional set of glasses and say, 'this abuse case that I deal with 100 times a week could in fact also be a CSEC matter. (Federal Law Enforcement, County 1)

One team member recalled after implementing the MDT, "We didn't deal with them as CSEC cases, they were dealt more as abuse/neglect cases… If I was an adolescent worker way back when and it was a runaway, which I honestly at the time probably couldn't have identified it" (Child Welfare Supervisor, County 2). Understanding CSEC as unacknowledged exploitation rather than a new phenomenon helped MDT members improve identification within their own agencies. The collaborative team process also helped foster the development of shared definitions of the phenomenon and its indicators among team members.

4.2 *Collaboration and Information Sharing*

The MDT approach not only increased local awareness about the problem but provided a forum to begin developing strategies to improve the response to the problem. Central to improving the response was enhancing collaboration and information sharing. Through collaboration, MDT team members began to recognize that CSEC cases were more complex than many traditional child abuse situations and required a different approach. As social workers within the state child welfare agency noted "I think that it [awareness] came because we all saw our part of the MDT. I think there were some cases where we were saying, 'what the heck is going on with this kid? They are on the internet, they are on Backpage.' We were kind of aware of it, but I think the training really started us thinking on what to, how as a community and what do we do and how do we even start this MDT team" (Social Worker, County 6).

Developing a multidisciplinary team response to CSEC builds cooperation between agencies and enhances the working relationships across public service sectors. Such collaboration improves law enforcement responses to CSEC victims and promotes better investigations. In describing the MDT work on an early case, one detective working in a county-based MDT for CSEC explained how the team approach changed their response.

> So this was a case that ran all the way through the night and into the next day, but what worked really well, even though we didn't have protocol yet at that point, we just knew who to contact. Contact the DA's office, contact the CAC, the CAC started the coordination process, got everybody involved. Of course, we had the police out in the field with us, and what was nice about it, which wouldn't necessarily happened before is we had a meeting within twenty-four hours. We had follow-up meetings within a couple of weeks. The email communication was great. Definitely some holes, like we have to do what we have to do very quickly…we talked about well maybe we don't want to rush to the forensic interview next time, maybe we want to do a couple of other steps. (Local Law Enforcement, County 2)

Strengthening communication and sharing information allows the team members from various agencies to be more aware of how to address cases of CSEC as a team and understanding the roles of each member. The importance of sharing information and collaborating was echoed by a residential service provider in another county. "I think the strengths so far are that it is enhanced communication. That by far, I think, has worked wonders. I think everybody is working off the same kind of play sheet. Everybody kind of understands what's happening, what is the criteria, and these are the steps that need to be taken … But I think everybody has a better understanding and is working more collaboratively than we have in the past, and I think that has been helpful" (Residential Facility Supervisor, County 5).

MDT models further improved relationships between agencies, creating a support system for team members, including law enforcement to feel comfortable reaching out for help, asking questions, and sharing information. A DCF director expounded on their experience with developing relationships and collaborating in the MDT. "I now have a very different relationship with the prosecutor that I feel like I can call and text day or night… same with the advocacy center and same with the service providers and so I think it has really enhanced our working relationship on all issues, CSEC or not" (Child Welfare, County 1). Improving professional relationships, the MDT model created an environment for team members to feel comfortable to contact each other, share information, and respond to issues together. Without collaboration between team members, agencies were likely to operate as independent islands, rather than informing necessary agencies of possible

CSEC. One social service worker stressed how team collaboration ignored boundaries to respond to a CSE youth, "the FBI got tipped, she was from out east and it wasn't this 'Well it's not my office and it's your problem', it was really about 'Hey there's a girl, it's an issue'" (Child Welfare, County 1). The team helped overcome the problem of responsibility for responses being siloed within organizations that may not communicate. Too often that resulted in gaps in the provision of services or safety for CSEC victims.

Working closely within MDTs helped develop relationships necessary to work across agencies and jurisdiction and create a safety plan for youth. Teams collaborating and sharing information not only creates a safety net for young people, but it can also support criminal investigations. While discussing an example of employing the MDT in response to a CSEC case, a prosecutor in one of the MDT counties remarked, "recently with the Department of Youth Services apprehension officer… we pulled together an MDT team and we talked about how best to approach this, and uhm, [local] PD was there… child welfare came… so we were all kind of had our own little pieces about the situation" (Prosecutor, County 1). Child welfare, Department of Youth Services, the District Attorney, and law enforcement shared information as a team regarding case details, offender history, and the youth victim, which lead to identifying the offender from an adjacent investigation and offer services to the exploited youth.

Several sites highlighted improved communication between child services providers and law enforcement to better assist in investigations. At one study site, local law enforcement supported the MDT approach in investigations explaining, "county emergency response team where the prosecutor, child welfare and local law enforcement all quickly team up together to coordinate the best resource to get this case investigated immediately" (Local Law Enforcement, County 2). Cooperating in teams allows each agency to offer missing details, as discussed by the prosecutor, "…our communication with other team members, with youth services, child welfare, and the gang task force, we're getting so much more information and coming at these people from every angle" (District Attorney, County 1). MDT site interviews confirmed that collaboration between agencies that come into contact with CSEC cases is critical to effectively facilitate information sharing. Providers can offer support for team members, while also communicating information to help all levels of law enforcement with investigations.

At a very practical level, youth service providers work with law enforcement within the MDT by reporting youth who may be at risk of exploitation and missing from care. Law enforcement can then be more vigilant in looking for these missing youth while on patrol. A residential service provider described the importance of this collaboration:

> I think there's a lot more collaboration between providers and child welfare and law enforcement. So there have been times where we've notified law enforcement that we're getting a referral for a girl who is high risk for exploitation. If we call about her being on the run [missing from care], you guys really need to be looking for her. And the police have responded well to that. With child welfare, when they've made a referral to us and there is a risk or there is a concern about CSEC the planning around it has been a little tighter and better. (Residential Supervisor, County 5)

4.3 Improving Response to CSEC

As MDT members receive training and education on CSEC and begin to work collaboratively on individual cases, they started to develop team goals focused on improving outcomes and meeting the needs of victims. A CSEC coordinator contrasted her experience in the MDT with the response to CSEC in the county prior to the implementation of the team.

> I mean it was all about getting the bad guy off the street and there was zero focus on the victim. And that's where I've seen the biggest change. It took my hearing voices raising, to come in here and says what's going on, and then wanting to go do a raid and watching the forensic interview, and this girl doing a phenomenal job of identifying this pimp and this chain and this whole nine yards. And it was like who was going to run out of here fast enough to go get the bad guys. I was so concerned about the victim and there was just zero focus on the victim. (CSEC Coordinator and Associate Director, County 2)

As the CSEC coordinator acknowledges, without a sincere focus on the well-being of the victims, law enforcement and investigators, due to the nature of their role gave greater attention to the offender rather than the youth helping with the investigation. Several teams recognize the need to respond to CSEC in a different manner than their conventional approach with one site recognizing, "there was exploitation and we weren't seeing that they weren't our traditional cases of child sexual abuse and that's when we acknowledged or we realized we needed to develop a different approach" (Child Advocacy Center, County 3). One example of this change is that MDT members did not focus on securing a forensic interview with the youth immediately after identification. During meetings, teams shift attention to the immediate needs of CSE youth and creating safety for the youth. Reflecting on responses prior to implementing the MDT approach to CSEC, a prosecutor in one county reasoned, "maybe the… the interview shouldn't have been the priority. Maybe there should have been more focus on the mental health needs and the support in the community versus the investigative needs" (Prosecutor, County 1). Evolving as a team in their understanding and approach to CSEC, MDT members shifted their mindset to prioritize meeting the needs of youth and building supportive communities around the youth. A child welfare provider working in one county MDT noted:

> But looking at them and kind of doing our first taking care of the child. Do you need food? Do you need a shower? Do you need a change of clothes? And not being concerned if they take off, but just trying to immediately meet those basic needs and come from a point of let's make sure your protected, let's make sure your medically cleared. (Child Welfare, County 2)

By responding with the intention of providing services and help for youth, teams can focus on the unique nature of CSEC cases and employ safety plans that allow youth to recover before going forward with a forensic interview. In one county, a team member from the CAC supported primarily focusing on youth services:

> We acknowledged early on that these cases are not going to be handled the same way. They might not result like we were talking earlier with a forensic interview for example, it might just be a multidisciplinary team discussion about how we can get the children or young adults into a setting that's going to be productive and supportive for them. (Child Advocacy Center, County 3)

Recognizing the importance of this shifting approach, team members move towards an understanding that immediate responses and longer-term goals for CSEC youth may be different than traditional criminal investigations of child abuse cases.

Understanding the complexity of CSEC cases, teams focused much of their early work on creating protocols, developing internal committees, and securing buy-in from local leadership. Teams convened to develop plans for youth, and in one instance a team member remarked that they are, "just making sure that we're providing for the immediate needs and long-term needs of these victims really crisis sort of stabilization, but also long term mental health and um jobs and things of that nature for these girls" (Local advocate, County 1). As the MDT members developed protocols to facilitate information sharing and coordination and conducted trainings in the local community to increase CSEC identification, institutional and resource concerns began surfacing. MDTs encountered challenges finding appropriate services and residential placements for youth identified as CSEC victims. In the next section, we discuss some of the difficulties MDTs faced and identify challenges inherent to the MDT approach.

5 Challenges of the MDT Approach

Communication and information sharing improved as the MDTs became more established, however, as the team worked on cases competing and conflicting interests between participating agencies began to emerge. For example, the goals of law enforcement to identify and indict suspects was sometimes at odds with local agency policy and practice around information sharing. One CAC encountered problems while coordinating with federal prosecutors participating in the MDT when they requested information about clients. As the CAC coordinator notes:

> What we more recently run into in is the federal prosecutors' office wanting our database. It's like, wanting it. Wanting it. Wanting it. We really couldn't figure out why and it wasn't just something we wanted to hand over. Thank goodness the federal agent participating on the MDT sort of validated our reality and was really shocked that we were being asked for it and what turned out was he wanted to bring it back to this law enforcement group. They basically wanted to see if any of the alleged perpetrators that we had on our list matched you know anybody else they were looking for or had something on. That really felt to us like that's a fishing expedition…That really is violating the confidentiality of our clients, of our identified girls. (CAC coordinator, County 1)

As the coordinator mentions, team members face challenges of balancing supporting investigations while complying with legal mandates protecting the information of youth clients. This example illustrates how the goals of law enforcement were sometimes in conflict with the practices and rules governing the activity of service providers.

Coordinating team responses and meeting the needs of all team participants is more challenging in situations where there is an absence of leadership among the actors who are contributing to the MDT collaboration. MDTs gather representatives from various sectors, but they often do not have a designated leader. Absence of a leader in the group designated to make final decisions and directed team member activities, meant that participants had to "understand where people's roles and responsibilities lie" (Child welfare, County 6). Across the counties, different MDTs had members with more significant leadership roles, but without an individual or agency tasked with taking a leadership role in the team, members can continue to act in siloes and unnecessarily repeat services and processes. In another county, MDT members became discouraged when the prosecutor refused perusing a case for criminal prosecution, "I will say a challenge down here is that our prosecutor's office, in my opinion, is not well respected. They're not responsive, the communication with them is very poor and they do not prosecute. So that is a huge weakness" (Child welfare, County 3). Concerned by inadequate communication between the prosecutor and other MDT members, MDT members expressed concern that team members, particularly within law enforcement would become less engaged in the MDT process or pursuing investigations if they were perceived as not resulting in outcomes that held offenders accountable.

Creating smooth coordination between law enforcement and other members of the MDT was a challenge in some communities. Where challenges existed, they often involved law enforcement team members expressing reluctance about CSEC during the team process. A child welfare director from one county mentioned trying to convince local law enforcement to buy-in to the idea of treating youth as victims of CSEC and responding in collaboration with the MDT.

> It's been really difficult to get the local police department which, you know, they are also under-staffed and under-resourced, but the buy-in has not been there. We have certainly had better luck with our federal partners and state police. You know, knowing who to contact, because of her knowledge base, she has been able to tap in and getting state police and certainly getting you know, whether it's FBI homeland security to help us to stop. The local police here has been a real challenge, partly because I think that they are dealing with the whole host of issues that they do not yet see how they are connected to CSEC. (DCF Director, County 1)

With a shortage of resources and personnel, local police departments may have limited capacity, to undertake organizational and routine changes in their approach to CSEC. Challenges with police involvement in MDTs was mentioned by many team participants across multiple counties. MDT members in one study site agree that some law enforcement still need training. One respondent expressed "There's education to be done there because we tried to have them [law enforcement] all come. They tried to have a law enforcement day for the training on understanding and awareness of CSEC and I don't even know if anyone came, or they cancelled it, or if there was a small showing, I'm not sure" (Social Worker, County 3). Untrained law enforcement may continue to treat CSEC cases as involving "throwaway" or runaway youth without connecting them with appropriate care and support services. Even worse, youth may be arrested for prostitution or other activities, labeled as delinquent, and processed through the criminal justice system as offenders. Law enforcement team members also noted struggling with decisions about whether to detain youth or proceed with CSEC investigations, not necessarily because they did not accept that CSEC is a form of victimization, but because of concerns about the lack of appropriate services that may exist in communities to keep exploited youth safe. As one law enforcement team member explained:

> I think the team is doing a great job. I think they did a really good job of having a lot of different voices there and a lot of different agencies because you do see that this affects so many people who are in this field. We have our residential programs, we have the clinicians, we have the school, we have our attorneys, we have our judges, it goes on and on. But…I think the struggle is what do we do next? What do we do if a child is picked up? Are we arresting them, are we not arresting them? Where are we holding them, are we notifying families, are we taking pictures, you know those have been the concerns I think everyone is trying to figure out. (Law enforcement, County 3)

As illustrated by the quote above, MDTs can improve the identification and coordination of CSEC cases to improve responses to victims in local communities, but it is critical for communities to develop the appropriate services to meet the unique needs of CSEC victims once identified.

6 Discussion

As law enforcement is increasingly tasked with the responsibility of identifying and responding to the commercial sexual exploitation of children (CSEC), it is important to develop partnerships with networks of social welfare and service providing agencies that are in the best position to meet the complex needs of youth victims. Multidisciplinary teams have proven to be an effective method to enhance communication and coordinate activity across a range of partners who come from different institutional environments with different structures, mandates, and goals. One of the key elements of the multidisciplinary team is that it provides the opportunity for team members to learn about a phenomenon from a variety of perspectives and share information that is traditionally siloed within a particular agency or domain. This is especially important in the case of CSEC because those youth most vulnerable to exploitation often have histories of engagement with social service organizations and may already be familiar to law enforcement. Understanding the phenomenon of CSEC and sharing information particularly about children who are victimized or at high risk for victimization is critical to creating networks of safety and improving individual responses to young people.

While MDTs can improve the response of a community to CSEC they are not always easy to develop or sustain. MDTs are temporary groups that require support and resources from partner organizations and agencies. The work on the MDT may not always be perceived as central to the mission of law enforcement partners, and at times the approach agreed upon by the team may not advance the goals of law enforcement, especially when investigations must be slowed or victims will not be available to be interviewed due to safety or concern about trauma. While participation in an MDT can support ongoing investigations and enhance the ability of the police to gather information and build a criminal case, the arrest and prosecution of suspects is not the primary goal of most CSEC MDTs which are focused on youth safety and improving the responses to meet the needs of youth. Addressing the commercial sexual exploitation of children is a difficult task, but strong partnerships with agencies and organizations in a community that are involved in the lives of young people at risk for victimization, including law enforcement, are essential in promoting safety and improving how communities respond to CSEC.

References

Andretta, J. R., Woodland, M. H., Watkins, K. M., & Barnes, M. E. (2016). Towards the discreet identification of Commercial Sexual Exploitation of Children (CSEC) victims and individualized interventions: Science to practice. *Psychology, Public Policy, and Law, 22*(3), 260.

Aron, L., Zweig, J., & Newmark, L. (2006). *Comprehensive services for survivors of human trafficking: Findings from clients in three communities*. Retrieved from Justice Policy Center at Urban Institute.

Bai, Y., Wells, R., & Hillemeier, M. M. (2009). Coordination between child welfare agencies and mental health service providers, children's service use, and outcomes. *Child Abuse & Neglect, 33*(6), 372–381. https://doi.org/10.1016/j.chiabu.2008.10.004.

Bell, L. (2001). Patterns of interaction in multidisciplinary child protection teams in New Jersey. *Child Abuse & Neglect, 25*(1), 65–80.

Bjelland, H. F. (2017). Identifying human trafficking in Norway: A register-based study of cases, outcomes and police practices. *European Journal of Criminology, 14*(5), 522–542.

Bracewell, T. E. (2018). Multidisciplinary team involvement and prosecutorial decisions in child sexual abuse cases. *Child and Adolescent Social Work Journal, 35*(6), 567–576.

Bradford, A. (2005). *Legal Outcomes as Effectiveness Research of Forensic Interviewing at the Child Advocacy Center of East Alabama (CACEA)*. Master of Science, Auburn University.

Brink, F. W., Thackeray, J. D., Bridge, J. A., Letson, M. M., & Scribano, P. V. (2015). Child advocacy center multidisciplinary team decision and its association to child protective services outcomes. *Child Abuse & Neglect, 46*, 174–181.

Bross, D. C., Ballo, N., & Korfmacher, J. (2000). Client evaluation of a consultation team on crimes against children. *Child Abuse & Neglect, 24*(1), 71–84.

Bureau of Justice Assistance. (n.d.). *Office of Justice Programs*. Retrieved from https://www.bja.gov/default.aspx

Cheung, M., & Brandes, B. J. (2011). Enhancing treatment outcomes for male adolescents with sexual behavior problems: Interactions and interventions. *Journal of Family Violence, 26*(5), 387–401.

Cole, J. (2018). Structural, organizational, and interpersonal factors influencing interprofessional collaboration on sexual assault response teams. *Journal of Interpersonal Violence, 33*(17), 2682–2703. https://doi.org/10.1177/0886260516628809.

Cole, J., Sprang, G., Lee, R., & Cohen, J. (2016). The trauma of commercial sexual exploitation of youth: A comparison of CSE victims to sexual abuse victims in a clinical sample. *Journal of Interpersonal Violence, 31*(1), 122–146.

Cross, T. P., Finkelhor, D., & Ormrod, R. (2005). Police involvement in child protective services investigations: Literature review and secondary data analysis. *Child Maltreatment, 10*(3), 224–244.

Cross, T. P., Jones, L. M., Walsh, W. A., Simone, M., & Kolko, D. (2007). Child forensic interviewing in Children's Advocacy Centers: Empirical data on a practice model. *Child Abuse & Neglect, 31*(10), 1031–1052.

Decker, S. H., Greene, J. R., Webb, V., Rojek, J., McDevitt, J., Bynum, T., et al. (2005). Safety and security at special events: The case of the Salt Lake City Olympic games. *Security Journal, 18*(4), 65–74.

Dinsmore, J. (1992–1993). A safety net of bridges: Interagency coordination on child abuse investigations. *Part I. National Center for Prosecution of Child Abuse Update,* 5–6(12–1), 1–2.

Edwards, J. M., Iritani, B. J., & Hallfors, D. D. (2006). Prevalence and correlates of exchanging sex for drugs or money among adolescents in the United States. *Sexually Transmitted Infections, 82*(5), 354–358.

Epstein, R., & Edelman, P. (2014). *Blueprint: A multidisciplinary approach to the domestic sex trafficking of girls*. Retrieved from Washington, DC: Center on Poverty and Inequality, Georgetown Law School.

Faller, K. C., & Palusci, V. J. (2007). Children's advocacy centers: Do they lead to positive case outcomes? *Child Abuse & Neglect, 31*(10), 1021–1029.

Farrell, A., Dank, M., de Vries, I., Kafafian, M., Hughes, A., & Lockwood, S. (2019). Failing victims? Challenges of the police response to human trafficking. Criminology & Public Policy. 18 3: 649-673

Farrell, A., McDevitt, J., & Fahy, S. (2010). Where are all the victims? Understanding the determinants of official identification of human trafficking incidents. *Criminology & Public Policy, 9*(2), 201–233.

Farrell, A., Owens, C., & McDevitt, J. (2014). New laws but few cases: Understanding the challenges to the investigation and prosecution of human trafficking cases. *Crime, Law and Social Change, 61*(2), 139–168.

Farrell, A., & Pfeffer, R. (2014). How police find cases of human trafficking: Cultural blinders and organizational barriers. *The Annals of the American Academy of Political and Social Science, 653*, 46–64.

Farrell, A., Pfeffer, R., & Bright, K. (2015). Police perceptions of human trafficking. *Journal of Crime and Justice, 38*(3), 315–333.

Feng, J.-Y., Fetzer, S., Chen, Y.-W., Yeh, L., & Huang, M.-C. (2010). Multidisciplinary collaboration reporting child abuse: A grounded theory study. *International Journal of Nursing Studies, 47*(12), 1483–1490.

Fryer, G. E., Jr., Poland, J. E., Bross, D. C., & Krugman, R. D. (1988). The child protective service worker: A profile of needs, attitudes, and utilization of professional resources. *Child Abuse & Neglect, 12*(4), 481–490.

Goldbeck, L., Laib-Koehnemund, A., & Fegert, J. M. (2007). A randomized controlled trial of consensus-based child abuse case management. *Child Abuse & Neglect, 31*(9), 919–933.

Goldstein, J., & Griffin, E. (1993). The use of a physician-social worker team in the evaluation of child sexual abuse. *Journal of Child Sexual Abuse, 2*(2), 85–93.

Grubb, D., & Bennett, K. (2012). The readiness of local law enforcement to engage in US anti-trafficking efforts: An assessment of human trafficking training and awareness of local, county, and state law enforcement agencies in the state of Georgia. *Police Practice and Research, 13*(6), 487–500.

Herbert, J. L., & Bromfield, L. (2017). *Better together? A review of evidence for multi-disciplinary teams responding to physical and sexual child abuse*. Trauma, Violence, & Abuse, 1524838017697268.

Hochstadt, N. J., & Harwicke, N. J. (1985). How effective is the multidisciplinary approach? A follow-up study. *Child Abuse & Neglect, 9*(3), 365–372.

Hurlburt, M. S., Leslie, L. K., Landsverk, J., Barth, R. P., Burns, B. J., Gibbons, R. D., et al. (2004). Contextual predictors of mental health service use among children open to child welfare. *Archives of General Psychiatry, 61*(12), 1217–1224.

ICAC. (2017). *Internet crimes against children task force program*. Retrieved from https://www.icactaskforce.org/Pages/About%20Us.aspx

ILO. (2017). *2017 global estimates of child labour and modern slavery – Americas regional brief*. Retrieved from Geneva: International Labor Organization.

Irwin, N. (2017). Police officer understandings of human trafficking and awareness of anti-trafficking measures. *Policing: An International Journal of Police Strategies & Management, 40*(2), 291–305.

Jacobson, M. (2001). Child sexual abuse and the multidisciplinary team approach: Contradictions in practice. *Childhood, 8*(2), 231–250.

Joa, D., & Edelson, M. G. (2004). Legal outcomes for children who have been sexually abused: The impact of child abuse assessment center evaluations. *Child Maltreatment, 9*(3), 263–276.

Johnson, L. B. (2013). A qualitative study of communication among child advocacy multidisciplinary team members using a web-based case tracking system. *Journal of Technology in Human Services, 31*(4), 355–367.

Jones, L. M., Cross, T. P., Walsh, W. A., & Simone, M. (2007). Do Children's Advocacy Centers improve families' experiences of child sexual abuse investigations? *Child Abuse & Neglect, 31*(10), 1069–1085.

Justice for Victims of Trafficking Act of 2015, Pub. L. No. 114-22, 129 Stat. 227 Stat. 129 (2015).

Kolbo, J. R., & Strong, E. (1997). Multidisciplinary team approaches to the investigation and resolution of child abuse and neglect: A national survey. *Child Maltreatment, 2*(1), 61–72.

Lalayants, M. (2013). Multidisciplinary collaboration in child protective clinical consultations: Perceptions of best practices. *Journal of Public Child Welfare, 7*(3), 253–274.

Lalayants, M., & Epstein, I. (2005). Evaluating multidisciplinary child abuse and neglect teams: A research agenda. *Child Welfare, 84*(4), 433–458.

Lalayants, M., Epstein, I., & Adamy, D. (2011). Multidisciplinary consultation in child protection: A clinical data-mining evaluation. *International Journal of Social Welfare, 20*(2), 156–166.

Landers, M., McGrath, K., Johnson, M. H., Armstrong, M. I., & Dollard, N. (2017). Baseline characteristics of dependent youth who have been commercially sexually exploited: Findings from a specialized treatment program. *Journal of Child Sexual Abuse, 26*(6), 692–709.

Le, P. D., Ryan, N., Rosenstock, Y., & Goldmann, E. (2018). Health issues associated with commercial sexual exploitation and sex trafficking of children in the United States: A systematic review. *Behavioral Medicine, 44*(3), 219–233.

Love, H., Hussemann, J., Yu, L., McCoy, E., & Owens, C. (2018). *Justice in their own words: Perceptions and experiences of (in) justice among human trafficking survivors*. Washington, DC: Urban Institute.

Maguire, E. R., & Katz, C. M. (2002). Community policing, loose coupling, and sensemaking in American police agencies. *Justice Quarterly, 19*(3), 503–536.

Miller, A., & Rubin, D. (2009). The contribution of children's advocacy centers to felony prosecutions of child sexual abuse. *Child Abuse & Neglect, 33*(1), 12–18.

Moran, P., Jacobs, C., Bunn, A., & Bifulco, A. (2007). Multi-agency working: Implications for an early-intervention social work team. *Child & Family Social Work, 12*(2), 143–151.

Newman, B. S., Dannenfelser, P. L., & Pendleton, D. (2005). Child abuse investigations: Reasons for using child advocacy centers and suggestions for improvement. *Child and Adolescent Social Work Journal, 22*(2), 165–181.

Newton, P. J., Mulcahy, T. M., & Martin, S. E. (2008). *Finding victims of human trafficking*. Retrieved from NORC: University of Chicago.

O'Brien, J. E., Li, W., Givens, A., & Leibowitz, G. S. (2017). Domestic minor sex trafficking among adjudicated male youth: Prevalence and links to treatment. *Children and Youth Services Review, 82*, 392–399.

Office of Juvenile Justice and Delinquency Prevention (OJJDP). Departments of Commerce, Justice, and State, the Judiciary, and Related Agencies Appropriations Act, Pub. L. No. 105-119, 111 Stat. 2440 (1998).

Office of Juvenile Justice and Delinquency Prevention (OJJDP). (n.d.). *Office of Justice Programs*. Retrieved from https://www.ojjdp.gov/

Olson, D. E., Albertson, S., Brees, J., Cobb, A., Feliciano, L., Juergens, R., Ramker, G. F., & Bauer, R. (2002). *New approaches and techniques for examining and evaluating multi-jurisdictional drug task forces in Illinois*. Chicago, IL: Illinois Criminal Justice Information Authority.

Onyskiw, J. E., Harrison, M. J., Spady, D., & McConnan, L. (1999). Formative evaluation of a collaborative community-based child abuse prevention project. *Child Abuse & Neglect, 23*(11), 1069–1081.

Pence, D., & Wilson, C. (1994). *Team investigation of child sexual abuse: The uneasy alliance* (Vol. 6). Thousand Oaks: Sage.

Preventing Sex Trafficking and Strengthening Families Act of 2014, Pub. L. No. 113-183, 128 Stat. 1919 Stat. (2014).

Reid, J. A., Baglivio, M. T., Piquero, A. R., Greenwald, M. A., & Epps, N. (2017). Human trafficking of minors and childhood adversity in Florida. *American Journal of Public Health, 107*(2), 306–311.

Ruhstaller, T., Roe, H., Thürlimann, B., & Nicoll, J. J. (2006). The multidisciplinary meeting: An indispensable aid to communication between different specialities. *European Journal of Cancer, 42*(15), 2459–2462.

Scott, W. (1998). *Organizations: Natural, rational and open*. London: Prentice-Hall International.

Shared Hope. (2016). *Protected innocence challenge toolkit*. Retrieved from Vancouver, WA.

Skaff, L. F. (1988). Child maltreatment coordinating committees for effective service delivery. *Child Welfare, 67*(3), 217–230.

Smith, D. W., Witte, T. H., & Fricker-Elhai, A. E. (2006). Service outcomes in physical and sexual abuse cases: A comparison of child advocacy center-based and standard services. *Child Maltreatment, 11*(4), 354–360.

Smith, T. M. (2011). *Case studies of multidisciplinary child abuse case review teams and their leaders in children's advocacy centers in Pennsylvania*. Ph.D. doctoral dissertation, Indiana University of Pennsylvania, ProQuest Dissertations & Theses Global database, Ann Arbor, MI. (3453625).

Tjaden, P., & Anhalt, J. (1994). *The impact of joint law enforcement-child protective services investigations in child maltreatment cases*. Retrieved from Denver: Center for Policy Research.

Turner, G. C. (1997). *Evaluating the interdisciplinary team approach to child sexual abuse investigation: A retrospective chart review comparing multi-agency and interdisciplinary investigatory approaches*. (PhD), New York University, ProQuest Dissertations & Theses Global database. Ann Arbor: New York University.

Victims of Trafficking and Violence Protection Act (TVPA) of 2000, Pub. L. No. 106-386, 114 Stat. 1464S (2000).

Twomey, M. S., Jackson, G., Li, H., Marino, T., Melchior, L. A., Randolph, J. F., et al. (2010). The successes and challenges of seven multidisciplinary teams. *Journal of Elder Abuse & Neglect, 22*(3–4), 291–305. https://doi.org/10.1080/08946566.2010.490144.

U.S. Department of Justice (USDJ). Providing Resources, Officers, and Technology to Eradicate Cyber Threats to Our Children Act of 2008, Pub. L. No. 110–401, 122 Stat. 4229 (2008).

U.S. Department of Justice (USDJ). (2010). *The national strategy for child exploitation prevention and interdiction*. Retrieved from Washington, DC: http://toolkit.heatwatch.org/policy/files/USDOJ_2010__National_Strategy_to_Prevent_CSEC.pdf

Varma, S., Gillespie, S., McCracken, C., & Greenbaum, V. J. (2015). Characteristics of child commercial sexual exploitation and sex trafficking victims presenting for medical care in the United States. *Child Abuse & Neglect, 44*, 98–105.

Wolfteich, P., & Loggins, B. (2007). Evaluation of the children's advocacy center model: Efficiency, legal and revictimization outcomes. *Child and Adolescent Social Work Journal, 24*(4), 333–352. https://doi.org/10.1007/s10560-007-0087-8.

From the Front Line: The Lived-Reality of Police Helicopter Operations

Simon Bennett

Abstract This chapter, based on interviews with National Police Air Service (NPAS) pilots and tactical flight officers, reveals the lived reality of police helicopter operations in England and Wales, from witnessing acts of extreme violence (such as a beheading in north London) to searching for the remains of bodies atomised by locomotive impacts. Highly professional in their outlook and behaviour, the men and women of the NPAS push themselves and their equipment to the limit in their quest to safeguard the public and apprehend criminals. Interviewees claimed a link between teamwork-training and performance. Teamwork-training – known in aviation as crew resource management – was universally praised. It is recommended that the police service urgently considers the introduction of teamwork training for all officers. The police service is starved of resources. The NPAS needs better equipment – longer-ranged, more capable helicopters with state-of-the-art sensors and communications equipment and protection from small-arms fire. The State's primary mission is to ensure the safety and security of its citizens. As a matter of urgency, the government should divert monies from health-care and education to the police service. Integrity and courage should be recognised and rewarded.

Keywords National Police Air Service · Interviews · Lived-reality · Mission-effectiveness · Operational safety

1 Introduction

The National Police Air Service provides helicopter support to the 43 police forces of England and Wales and the British Transport Police from a network of 14 bases (National Police Air Service 2019). The NPAS was created in 2012 from a variety of providers, the intention being to improve efficiency and effectiveness. The NPAS

S. Bennett (✉)
Civil Safety and Security Unit (CSSU), University of Leicester, Leicester, England
e-mail: sab22@leicester.ac.uk

B. Fox et al. (eds.), *Science Informed Policing*, Advanced Sciences and Technologies for Security Applications,
https://doi.org/10.1007/978-3-030-41287-6_11

has been criticised on grounds of cost and service-delivery (Her Majesty's Inspectorate of Constabulary and Fire and Rescue Services 2017; Collins 2017; Bennett 2018). Presently, the NPAS is in a state of flux. Changes being considered include reducing the number of bases and introduction of fixed-wing aircraft in lieu of helicopters.

Creation of the NPAS coincided with deep cuts in police funding and multiplying operational demands. David Thompson, chief constable of the West Midlands force and the finance and resources lead for the National Police Chiefs' Council, claims UK policing is having to accommodate "steep budget reductions and a widening mission". According to Thompson: "There has been a real-term reduction of police budgets of 19% since 2010, but ranging between 11–25% across forces" (Thompson cited in Dodd 2018).

The author interviewed NPAS pilots and Tactical Flight Officers (TFOs) in 2017. In that year, the NPAS operated a mixed fleet of Eurocopter (EC) 135 and 145 twin-engine, four-seat helicopters (Fig. 1). Equipped with a variety of sensors (for example, an infra-red camera and night-vision goggles) and mapping systems, these small, agile machines were operated at altitudes of between 1000 and 1500 feet to visual flight rules (VFR). A key standard operating procedure (SOP) was not to fly in, or above cloud. Operating minima were:

- Cloud-base 300 feet (around 50 feet above the tops of electricity pylons)
- Daylight visibility 1 km
- Night visibility 3 km (night-vision goggles were available).

Fig. 1 A National Police Air Service EC135 at Gloucestershire Airport. (Wikimedia Commons 2016)

At the NPAS, work is organised around twelve-hour shifts, with handovers at 07:00 and 19:00. Usually, the incoming shift is briefed by the outgoing shift (if not all, then some of the incoming shift are briefed). The NPAS synthesises two very different work cultures – aviation and policing. NPAS pilots work to European Aviation Safety Agency (EASA) flight-time limitations (FTLs). TFOs work to police shift-work rules. Serving with the NPAS transforms police officers' outlook. As one TFO put it: "[One's] police officer … identity is taken away massively. You become aircrew". Pilots and TFOs work a four-days-on, four-days-off pattern. One TFO remarked that the first rest-day is spent catching up on sleep. Sleep-debt reduces performance.

The EC135/45 usually carries a crew of three: a pilot and two TFOs. The TFO in the front left-hand seat operates the camera and handles some of the radio traffic. Tactical control of the sortie is vested in the TFO in the rear right-hand seat. This TFO operates the mapping system using a keyboard, accepts or declines tasks (in consultation with the other TFO and pilot) and manages the bulk of the communication with ground units and Flight Dispatch (located in a purpose-built suite in a West Yorkshire Constabulary facility in Wakefield). During single-TFO sorties, the TFO will occupy the rear right-hand seat. A pilot commented that non-verbal communication, which accounts for *circa* 55% of communication (Fig. 2), is impossible in this configuration. Tandem seating inhibits communication.

On-board duties are not determined by rank. It is possible that a PC will occupy the rear right-hand seat (from which s/he will exercise tactical control of the sortie) and a sergeant the front left-hand seat. A TFO observed: "[I]f you are in the back … you have the shout". Another said: "The pilot does not decide on the tasking. The pilot is a taxi-driver, effectively. They go where we ask them, if it is within limits".

To gain a better understanding of the missions, the author occupied the rear left-hand seat (and, on the occasion of a ferry-flight, the front left-hand seat) on 18 sorties. These observation flights, undertaken *post* the interviews, revealed the lived reality of police helicopter operations. Noteworthy aspects included:

Fig. 2 Elements of the communication process. (Monarch Airlines 2017)

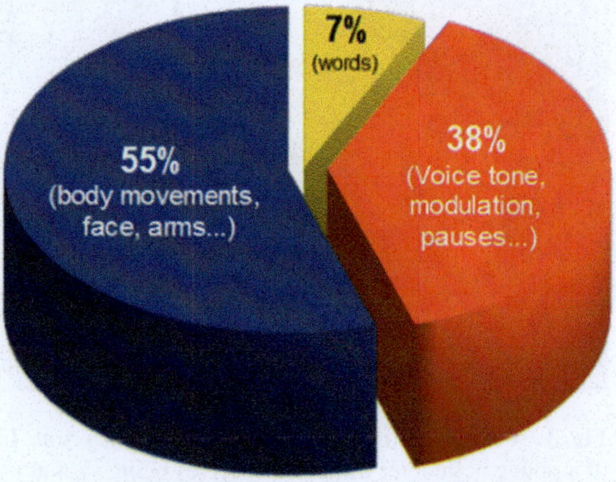

- the operational imperative to launch as quickly as possible once a task had been accepted
- operating at low altitude in uncontrolled airspace in proximity to other aircraft and birds. Helicopters are vulnerable to bird-strikes
- transiting controlled and uncontrolled airspace at close to maximum speed (known to the crews as straight-lining) to arrive on-scene as quickly as possible
- the need to create a mental model of the incident, task and options from multiple reports of varying quality submitted across busy communications channels in a noisy, vibrating cabin
- crewmembers' ability to quickly develop robust work-arounds for technical faults (such as communications breakdowns). Pilots and TFOs use *bricolage*, a problem-solving technique grounded in self-reliance, creativity and making-do (Sanchez-Burks et al. 2015; Phillimore et al. 2018)
- the mental capacity and stamina to accommodate no-notice, *en-route* re-tasking. A crew could be re-tasked multiple times
- the mental capacity and stamina to multi-task. For example, crewmembers are required to monitor six police, and two air-traffic channels while scanning the ground for suspects, communicating with one another and making tactical decisions
- the need to maintain a high state of situation awareness throughout the sortie, paying close attention to weather, fuel-state, potential air-traffic conflicts, terrain issues and threats from miscreants (helicopters are shot at and are subjected to laser attack)
- the need to perform when faced with traumatic or threatening situations, such as witnessing acts of extreme violence, following a railway track to locate a person hit by a train, being shot at (Fig. 3) or being targeted with a laser

Fig. 3 A shop set alight during the 2011 English riots. In Birmingham, on 9 August 2011, the crew of a police helicopter filmed themselves being shot at by a rioter. (Wikimedia Commons 2011)

- the need for the pilot and TFOs to maintain the highest standard of teamwork regardless of the number, severity and duration of threats and their own physical and psychological state (Bennett 2018).

In 2017, the author conducted semi-structured interviews (see questions at Appendix) with four pilots and eight TFOs at several NPAS bases (the NPAS pilot responsible for the service's CRM programme monitored the first interview). The interviews were conducted in private. Responses were tape-recorded and transcribed verbatim by the author. The transcribed interviews ran to over 30,000 words. Narratives were subjected to textual analysis (TA) and content analysis (CA) (Dixon et al. 1987; Taylor et al. 1995; Neuman 1997; Maxfield and Babbie 2001; Braun and Clarke 2006), enabling identification and substantiation of themes and trends, which are described below. As mentioned, the author also observed 18 sorties (Fig. 4).

Fig. 4 Straight-lining to a job in the Midlands in an EC135. Bird-strikes are an ever-present threat. (Author's picture)

2 Themes Identified from the Interview Data

2.1 *Resourcing Affects Teamwork, Safety and Operational Effectiveness*

Police budgets have been cut. Some claim the cuts have had a detrimental effect on front-line policing. In late 2018, the Chief Constable of the West Midlands force observed: "The public's experience is policing that is less visible, less responsive and less proactive .... Core aspects of policing – such as answering calls, attending emergencies, investigating crime, bringing offenders to justice and neighbourhood policing – are being pushed beyond sustainability, and are in danger of becoming ineffective, to the detriment of confidence in the police" (Thompson cited in Dodd 2018). In its 25 October, 2018 report *Policing for the future*, the House of Commons Home Affairs Committee (House of Commons Home Affairs Committee 2018) observed: "Policing is struggling to cope in the face of changing and rising crimes, as a result of falling staff numbers, outdated technology, capabilities and structures, and fragmented leadership and direction. Without significant reform and investment, communities will be increasingly let down".

Some interviewees claimed that budget cuts have caused the NPAS to change its *modus operandi*. For example, dispatching aircraft with one TFO only. A Base Commander-TFO claimed that single-TFO sorties added risk: "The systems in the aircraft are configured for three-person operation. The systems are not compatible with two-person operation. You don't get the full benefit of the systems. If you are on a slow-time job – for example, missing from home in a rural area – you can do it with two people. Albeit you would not get the result you would get with three people. Fifty percent of our searching is done with the camera. The other fifty by the rest of the crew with eyes-out. With a two-crew aircraft, you have the pilot who is flying the aircraft and glancing out, and the person on the camera, who is eyes-in, working the camera. They are also having to work out where they are and be situationally-aware. So they are not giving full commitment to looking at the screen and searching, because they are having to look where they are, switch it to the mapping system to work out the location of possible sightings, use the radio to talk to people on the ground whilst monitoring the dispatch channel and air traffic channel. You can get into max-overload very quickly, even on a slow-time job. On a high-demand job, such as a firearms incident, you cannot give a proper service. I'd say you are putting yourself and that aircraft at higher risk because you are maxing-out so quickly with your workloads. Now, I understand why they want two crew ... because we only have so many aircraft and we have to provide a service. Most bases will offer two crew, but they'll say we can't do that job because I'm not competent to deal with that job. The big danger is that you are tasked to a slow-time missing-from-home, and while you are on that job, you get something else. If you turn it down it damages the organisation's reputation. Also it puts pressure on the individuals to say 'yes' and to try and provide a service".

A pilot noted that the NPAS trains for three-crew operation (pilot and two TFOs). He claimed that the tandem seating arrangement dictated by single-TFO operations inhibited non-verbal communication, and that pilots' natural desire to support colleagues could see them drawn into the policing side to the detriment of flight-safety: "Normally, you train as a crew of three. When you single-crew, you no longer have the cross-cockpit non-verbal communication with the TFO. You can't poke him [sic] in the shoulder to get his attention, because he is now sat in the back. Because the guy in the back is now working like a one-armed paper-hanger, it is only natural that me, as the pilot, wants to help him out. The problem then is that you get sucked into the job [policing] side of it, and pay less attention to … not flying into anything. I think it is quite a big CRM issue. We are not trained for single-crewing. No one says: 'When you are single crewed, this is what you have to do'". Single-TFO operations could be classed as a latent error, as defined by Reason (1990, 1997, 2013).

A Base Commander-TFO described how staff shortages impacted capacity: "I have enough staff to crew the aircraft every day, so long as everybody is here. If people take leave it gets trickier. The shift-pattern is designed for twelve people, and I have eleven. It's a balancing act". According to the Base Commander-TFO, complaints had fallen on deaf ears: "Decisions are made in Wakefield without much consultation. For the last two or three years the understaffing issue has been flagged up by the base managers. It has been flagged to the senior leadership team and to HR [Human Resources], but nothing has been done about it".

2.2 *Within the NPAS, Perceptions and Behaviour Are Influenced by Sub-Cultures, Both Functional and Dysfunctional*

Organisations, consisting of people, goals, processes/procedures, culture, technology and buildings/infrastructure, operate within a context that is shaped by regulation, economic circumstance and stakeholder demands (Davis et al. 2014). Sub-cultures, associated with interests, beliefs, skills, gender, salary, career-stage or physical location (Turner 1988; Kilmann 1989; Trice 1993; Brown 1995; Hatch 1997), are a feature of organisational life (Eyre 1984). Sub-cultures may induce subscribers to undermine the host (Armstrong 1996). Employees may express frustration or disaffection through "organisational misbehaviour" (Watson 2003). Organisational misbehaviour can occur at any level, from the shop-floor to the boardroom (Watson 2003). Employees may associate themselves with, and reproduce/amplify, more than one sub-culture (Waring 1996).

Some interviewees claimed that a strained relationship existed between the bases and the national organisation. According to the Base Commander-TFO of a base in the south of England, the act of absorbing numerous force or regional helicopter support units, with their distinctive traditions and aviation cultures, into a national organisation whose senior managers were regular police officers, had stressed the NPAS.

The cultures of aviation and policing are very different. Aviators, embedded in close-knit teams and indoctrinated with teamworking protocols such as crew resource management (CRM), are trained, encouraged and expected to challenge decisions that, from a safety or efficiency perspective, they feel uncomfortable with (Kanki et al. 2019). Police officers, who receive no formal training in teamwork, are less inclined to challenge decisions or admit to errors. A TFO explained: "As a police officer, you don't get any teamwork training. I started my CRM training pre-NPAS. It was completely eye-opening that we should challenge regardless of rank or seniority. The police had taught me the opposite: that you should just obey and take orders. CRM taught me that I should not be intimidated by rank or qualifications: that the point is to get home safely. That you should not blindly accept what is going on". Another TFO said: "As a police officer, CRM is completely new to me. We don't have anything like that in the Police. The culture within the Police is very much … depending on where you work … if you make a cockup, don't tell anyone. CRM is very much the opposite. There is a big change for police officers new to this aviation environment. It was a big change … understanding the culture of CRM and what it means". The Base Commander-TFO of a northern base said: "You can see the difference … between standard policing and police aviation. In the former … there isn't the openness and honestness [sic] around those teams, and people tend to be told to do things, and they do them, and there is a fear of challenging". The Base Commander-TFO described the relationship between the bases and headquarters in these terms: "We are an aviation business involved in policing. However, at the top we are policing with aviation involved …. There is still too much of an attachment to West Yorkshire Police". A TFO observed: "The whole idea of being able to challenge your manager in the right circumstances should go across the Police …. If you have a better idea, they [senior officers] should be receptive …. If something needs saying, it needs saying. I think I can do that in this organisation [the NPAS]. If I couldn't … I wouldn't be happy working here".

Some interviewees claimed that locating the NPAS within the institutional setting of the West Yorkshire Constabulary had led to friction. The Base Commander-TFO of a base in the south of England said: "The top brass does not have an aviation or an air support background. They have an operational policing background. Probably, to start off with, they were not used to being challenged in the way that I would expect to be challenged in the aviation world". In terms of safety, a propensity to challenge is a positive. The Base Commander-TFO claimed the culture and politics of the NPAS may lead employees to self-censor: "There have been quite a few occasions over the last year or eighteen months where the hierarchy of the organisation has really put the fear of God into the middle management. Which left people either fighting the organisation, or shutting up and hunkering down …. Sometimes when people are jaded, rather than find a positive way to develop a new process, they just wait for it to fall over. There is some oppositional behaviour between some people in our organisation and the top brass". A TFO observed of senior officers' behaviour: "They are resistant to change. Their way is the right way. 'I am a senior officer, I am right'".

2.3 *Generally, Frustrations Are Met with Circumspection and Resolve*

There was universal agreement that safety came first and that the frustrations of the job should be viewed through a safety-first prism. As a TFO put it: "You have got to be a bit stoic and say: 'We'll get them next time'. 'It will happen again. We'll make a difference, more often than not'. You kind of pacify yourself that way. You think of all the times you have made a difference. You always get them in the end. Because they will always do it again. And you get them eventually. They [the perpetrators] don't know what else to do".

2.4 *The Teamworking Standard CRM, Which Is Trained by the NPAS, Provides Practical Support and Emotional Comfort in Challenging Situations*

National Police Air Service pilots and TFOs are jointly trained in the teamworking standard CRM. Crew resource management instils in flight crew the skills and confidence required to assemble, organise and direct resources (hardware, software and liveware) for safe and efficient flight. CRM draws on various academic traditions, including Trist and Bamforth's (Trist and Bamforth 1951) socio-technical systems theory, Tuckman and Jensen's (Tuckman and Jensen 1977) theory of work-group formation and Langer's (Langer 1989) theory of mindfulness. Trist and Bamforth argued that work teams perform best when they are required, equipped and empowered to see a task through from beginning to end. Crew resource management training aims to improve listening, observational, planning, co-ordination, leadership and teamworking skills. Trainees are encouraged to mine for ideas, ask questions and innovate. CRM training makes the case for methodical scepticism and constructive criticism (Wiener et al. 1993; Green et al. 1996; Krause 1996; McAllister 1997; Maurino et al. 1998; Campbell and Bagshaw 1999; MacLeod 2005; Gordon et al. 2013; Clay-Williams et al. 2014; Civil Aviation Authority 2016a, b; Bennett 2019a).

It was universally held that the NPAS's CRM training programme helped pilots and TFOs develop the skills and attitudes required for teamwork. Some claimed that their CRM training had helped them work through stressful situations. One TFO who, as camera operator, had witnessed a beheading, observed: "You tend to lean on it [CRM] sometimes. I had quite a horrific incident a couple of years ago. We were dealing with a call to a chap with a knife in North London [Edmonton] .... It quickly developed into something very significant where a chap who turned out to have mental issues actually beheaded somebody on camera while we were watching. He held her head up to us. Threw it down the garden. You don't expect to see that. He continued down a row of gardens towards children playing. He had a bloodied knife in one hand. You are thinking 'Here we go'. As it was going on, and we were all using colourful language … myself especially … we were all almost leaning on CRM as a kind of a support: 'OK boys, you've got to remember what you are doing'. 'Pilot, are you happy? Are you safe? Are you level? Are you with it? Are you OK?' 'Guy in the back. Andy. How are you doing?' Because Andy's brother was actually down there dealing with this guy as well. The radios were dropping out. The mapping was inaccurate. So there was a lot of pressure. And we kind of almost took ourselves out of it by using that interaction, not as a crutch, but as a guide to get us through the situation, so you almost made a point of verbalising your thoughts more than you would normally: 'OK. We've got to think about the SOPs for landing'. It was really useful to lean on the awareness of CRM and things like the SOPs. We were making sure we were talking [verbalising], which is what CRM is all about: 'Is the aircraft good?' 'Are you good?' 'Are we ready to do the pre-landers? [Prior to landing, the TFO in the front left-hand seat runs through a check-list with the pilot]'. 'Is everyone happy that we have done the pre-landers correctly?' 'Do we need to do them again, because we have all had one Hell of a shock?' There was a lot more chatter and interaction than there would have been without awareness of it [CRM]".

Asked if this was the worst incident he had attended, the TFO said: "I'd seen equally messy things … train strikes … people being shot on camera … and then the live stuff … you go to a road traffic accident and you've got an old dear caught by a bus wheel and she is seven feet long … but this was the first time I'd really seen anything quite so dynamic and adversarial, if you like, caused by one person deliberately to another, so it was probably the worst thing I have ever seen. So yes. Quite a shocker, really".

The stress inherent in this incident was exacerbated by circumstance. Specifically:

- The crew had been tasked to this incident late in a sortie. Instead of landing back at base, the crew found itself catapulted into a major incident. As the TFO explained: "[It was] one of those jobs where you are on your way home, you can see your landing site, and it just came over the radio, it was one of those very low-key things. It quickly developed into something very significant … "
- The TFO in the rear right-hand seat was related to one of the Metropolitan Police officers who attended the incident. Because there were no Taser-carrying officers or armed response units available during the early stages of the incident, first-responders were in jeopardy. The perpetrator, Nicholas Salvador, was a paranoid schizophrenic who smoked skunk cannabis, claimed by some medical practitioners to induce psychosis (British Broadcasting Corporation 2015).

During another incident, CRM helped a crewmember identify and resolve a breakdown in teamwork and functionality. As the Base Commander-TFO of a base in the south of England explained: "[O]ne of my crews attended where a horse was butchered during a riot at a town near here. On that occasion, the crew was unable to intervene and prevent that from happening. At the time, the performance of one of the crewmembers in the cab deteriorated very badly, and the other person involved was able to recognise that and take on the communications aspect of their role as well. There we had two people. One with good CRM skills who used them. One with poorer CRM skills who was struggling. I don't think in their case it was a lack of CRM training or input that caused the problem. It was down to a more general difficulty of dealing with stress and maintaining situational awareness in stressful situations".

In another incident, CRM helped the TFOs and pilot maintain focus and deliver a result during a pressured search for a missing person performed in challenging conditions. As the TFO who occupied the rear right-hand seat explained: "About three years ago [2013], two teenage girls went swimming in the River Wear. They went under. The police retrieved one body. The other body could not be found. I was in the back. We did a sweep of the river. We used the searchlight. I was lying on the floor of the cabin looking out the door. Murky water. The pilot said we're getting low on fuel and we are losing the light. I went on the radio and said: 'This is probably going to be our last sweep'. You could feel the tension from the Control Room. Basically, the police wanted to say: 'Here's your daughter'. There was a lot of pressure on us. People on the ground were not having any luck. We eventually found the body and talked officers on to the location. We performed well as a team. It was like a very sombre victory. We did it methodically. There was pressure and upset. CRM helped us through it". Asked to reflect on *how* CRM helps in charged situations, the TFO said: "It makes you look out at yourself. It's as if you can look in and see yourself: 'I know what is happening here'. It is like me looking down, and seeing how we operate. Seeing how this can go really bad if we are not careful: Look at it; Reassess; Talk about it".

This testimony suggests the CRM skillset helps pilots and TFOs cope with challenging situations. As a coping-mechanism, CRM helps in several ways:

- Crew resource management training promotes mindfulness (see Langer's 1989 definition) and high-reliability operating (see Mason's 2004 definition). Potentially, it equips trainees with a capacity for calm, unhurried reflection. In an operational context, this capacity can deliver important benefits, including: accommodation of distractions and stressors; identification of weaknesses; canvassing of opinion; formulation of effective action plans. Accommodation of distractions and stressors helps subjects maintain focus. Identification of weaknesses (for example, an underperforming colleague) supports effective working. Formulating action plans keeps minds focused by structuring and directing activity.
- Crew resource management encourages the sharing of information and ideas. It lays the foundations for synergy. Interviewees linked the volume and quality of communication with mission success. The TFO who witnessed the beheading observed: "CRM is about everyone constantly bouncing off everyone else. It worked. It really worked". Effective CRM supports high-reliability operating: "*[Highly reliable] organisations* are constantly aware of the possibility of failure,

appreciate the complexity of the world they face, concentrate on day-to-day operations and the little things, respond quickly to incipient problems and *accord deep respect to the expertise of their members. They value knowledge and expertise highly [and] communicate openly and transparently* [my emphasis]" (Mason 2004).

2.5 CRM Breaks Down Occasionally

When CRM fails, the consequences can be serious. In 1990, a Boeing 707 operated by Colombian airline Avianca crashed due to fuel exhaustion. Reasons for the Flight 52 mishap included the crew's inadequate CRM (non-technical) skills. Crewmembers failed to monitor and cross-check colleagues' statements and actions, and failed to provide air traffic control with an accurate picture of the aircraft's fuel-state. There were 73 fatalities out of 158 passengers and crew (Brookes 1992; Federal Aviation Administration 2017).

Some interviewees claimed to have witnessed breakdowns in CRM. A TFO recalled a sortie occasioned by a maritime emergency: "We went to look for two divers who were lost off St Mary's Light House [Tyne-and-Wear, North East England]. The pilot decided that when the SAR [Search and Rescue] helicopter arrived on scene, we were going. Normally we would split. SAR would go low, and we'd go high, to put a safety margin in. The pilot insisted on leaving the scene. I questioned his decision. He said: 'I'm the captain of the aircraft. We're going'. We got back and had a big row. I politely said that there were things we could have done on-scene. He said: 'I'm not going to discuss this with you'. I said: 'Fair enough … but I'm going to tell you'. Like the NHS, my job is vocational. I'm here to help people. The day you stop wanting to help, leave. He was a civilian pilot who wasn't bothered what was going on in the water. Me and my colleague in the aircraft were wanting to help. They were found a few hours later, dead. Both drowned".

Another TFO recalled an occasion when crewmembers disagreed over the feasibility of a task: "We were flying to a job on a night when the weather wasn't very good. Bravado kicked in a little bit. We were tasked to a job close to the coast. There was a bank of cloud. We said that we should turn the job down. The pilot said: 'Stop being soft. Man-up. We can't turn down all jobs because of weather'. That gets your hackles up. This is CRM breaking down. Totally wrong attitude. We are not allowed to fly in cloud. We went to the job. It wasn't a particularly good job. We didn't need to risk ourselves for it. We turned around to look at the airport and it was covered in cloud. We had to fly above the cloud, which you are not supposed to do. We can't get into the airfield. A small hole opened up. We didn't know what was through that hole. We flew through that hole. We could have hit power lines. Very, very dangerous. Managed to get in. The pilot admitted he'd screwed up. I was angry because we'd risked our lives because of bravado".

A third TFO recalled working with a pilot whose behaviour was far from mindful: "When I came to the Unit … you had a Chief Pilot … who was old-school, ex-military. There is no such thing as a stupid question. But if you asked something that you think is a sensible question, you could get rebuffed. 'What you asking that for?'"

A Base Commander-TFO recalled a breakdown in CRM at a critical moment: "[One of our TFOs] has a tendency to move into a state of auditory exclusion .... [T]hey were deeply involved in a task. They had got to that stage of auditory exclusion, and they are shushing the pilot as the pilot calls up a 'close-by' TCAS [Traffic Alert and Collision Avoidance System – a device that scans for aerial threats] hit". A TCAS hit requires the pilot and two TFOs to visually scan for potential traffic conflicts. Crewmembers reviewed the event at the end of the sortie: "Everyone in the cockpit was able to recognise what was going on .... [A]t the time [CRM] did not get us out of trouble, but it gave us a framework to address a problem".

A pilot recalled an occasion when his crew voiced concerns about a mission, but he decided to press on: "We were going to a job. The crew were aware of the minima, but they had not experienced them. There was some bad weather coming through. We had 2,000 metres visibility. The cloud-base was about 500 ft. We were following a railway line. I told them that I knew the area. I knew there were no power lines or high obstacles. I said: 'I'm quite happy to push on'. But they were still unsure. They were saying 'We're not happy with this'. But they listened to me. It was just a thin band of weather. We got through. Now they have experienced it".

2.6 *The More Challenging and Visceral the Task, the More Difficult the Post-Sortie Adjustment*

Several TFOs and pilots talked about the emotional strain of high-consequence, time-critical jobs, such as searching for a person swept out to sea, or a person struggling to stay afloat in a fast-flowing river. Some jobs were so out-of-the-ordinary that they took on a surreal quality. The TFO who witnessed the Edmonton beheading recalled sitting with his colleagues, post-sortie, in a police canteen in what can only be described as a state of bewilderment: "We were all just in the canteen saying: 'Can you believe that?' 'What the Frigging Hell just happened?' 'Did that really happen?'" Asked to describe his feelings on witnessing the beheading, the TFO said: "I felt sick. Kind of cold. Very, very stunned". The TFO who searched for the missing diver felt aggrieved when his pilot abandoned the search. He confronted the pilot post-sortie.

Wikler's (1990) observations on American soldiers' responses to being transported in a matter of hours from a battlefield in Vietnam to their home-town are pertinent: "Not uncommonly, a soldier … [will] travel fifteen hours … arrive on a base and six to eight hours later be on the street as a civilian. Such procedure disallowed the possibility of veterans collectively 'working through' their war experiences, and possibly resolving or coming to terms with the injuries of self". Many of the young men who fought in Vietnam experienced post-traumatic stress disorder

(PTSD) (Card 1987; Barrett et al. 1996). During the Second World War, troops returning to the US from Europe and the Far East were carried on troop ships, allowing time for reflection, adjustment and decompression. More recently, the British Army arranged decompression stopovers for soldiers returning to the UK from active service in Afghanistan and Iraq. Soldiers were staged through Cyprus, where they were rostered a few days rest and relaxation. Psychological support was provided.

While the crew that witnessed the beheading was immediately stood down, others found themselves preparing for the next job with no time set aside for reflection, adjustment and decompression. To borrow Wikler's phrase, often there is no time available to 'work through' recent, potentially traumatic events. Frequently, pilots and TFOs are catapulted from one challenging, stressful job to the next.

As mentioned, the crew who witnessed the beheading struggled to make sense of what they had seen. Symptoms of PTSD include re-experiencing the traumatic event and physiological reactivity (Christie and Evans-Pim 2012). Those with PTSD experience "recurrent and intrusive distressing recollections of the [traumatic] event, including images, thoughts or perceptions" and may act or feel "as if the traumatic event were recurring … includes a sense of reliving the experience … and dissociative flashback episodes" (MacNair 2012). According to Hodgkinson and Stewart (Hodgkinson and Stewart 1991) a person's capacity to successfully process a traumatic event is influenced by:

- The subject's mental state at the time the traumatic event was witnessed
- The nature of the traumatic event itself (for example, whether such an event could reasonably have been expected to occur)
- The timing and quality of post-event trauma counselling.

The crew that witnessed the Edmonton beheading received counselling. The TFO who had been on the camera described how events unfolded post-mission: "We basically got mugged by my sergeant. He said: 'Right. Don't touch anything. Out of the aircraft. Canteen. Sit down. Just, you know, let it sink in. Calm yourselves down'. Then they started the post-trauma thing. There is a procedure they run through, and we got counselling. I think I had two or three counselling sessions. I immediately went on to some leave. That was my last day before a week off. So that kind of helped. I just went off down to the south coast for a week. Then we kind of got together about a week or so later on, and just sort of, you know, asked 'How are you doing?'. Obviously, we were in contact while we were off. There was a little outpouring of emotion".

2.7 CRM Helps Pilots and TFOs Manage Stressors

Police helicopter operations are suffused with stressors. For example: pressure to launch quickly; operating at low altitude in marginal weather; vibration; noise; pressure to turn the aircraft around quickly; the frustrations of inactivity; irregular meals; sleep-loss; disturbed sleep.

Sleep-loss affects performance, which deteriorates during the two circadian lows (Bennett 2010). One TFO reflected on how sleep deprivation affected his mood and performance: "After 02 [2 a.m.] I am quite grumpy. You are heading for the 04 [4 a.m.] trough …. It is dark, your body is in a different state. On a night shift, I go from sleeping seven hours to sleeping two. You are robbed of sleep. Doing four hours flying on a shift is quite a long time. You get back, you get gassed up. Do your admin. You end up throwing your food down your neck …. Three hours [flying] makes an impact on you. My ears are ringing sometimes. The aircraft vibrates. It just tires you out. You might be bouncing around. Your eyes get quite strained".

Another TFO reflected on the frustrations of down-time and working four consecutive, twelve-hour shifts: "People can get agitated and irritated while on the ground …. We get along by ribbing and taking the mickey out of one another. But sometimes it can be taken too far". Several interviewees claimed the CRM skillset helped them manage stressors. A pilot said: "CRM does flag up where things are about to go wrong …. If things are getting kind of hectic, you feel a bit more confident in saying 'Stop'. That is … a big thing". Crew resource management, which is informed by HRO theory (see Mason's 2004 definition), promotes mindful behaviour.

2.8 Expanding the CRM Training Catchment Would Boost Efficiency and Effectiveness

It was held that expanding the CRM training catchment to include Dispatchers, managers and others would help deliver a more effective service. A pilot noted: "[Control room staff] are not hugely aviation-aware". Asked who should be trained in CRM, a TFO observed: "The Wakefield Dispatchers … regional managers, whether they come from a flying background or not …. Senior people need to spend time with us. They need to understand the basics and the demands …. One time in a twelve-hour shift I did five hours and fifteen minutes flying. You might say 'That's not even half'. You are absolutely knackered after four hours. It's not like sitting on the flight deck at 35,000 feet on autopilot, monitoring".

A Base Commander-TFO observed: "I think [Dispatchers] should be attached to a Unit for a week or two to gain a better understanding of the working environment and the demands …. They need to see what happens in the aircraft and what happens on the ground between jobs. They need a better understanding of the paperwork side of down-time. They can sometimes be too quick to send you to another job. They need to understand that when you land … after that second job, you have twice as much paperwork to do …. [T]hey should see CRM practiced at a Unit level".

Another pilot observed: "When a Unit is running CRM training, ops staff and managers should attend …. CRM is much better when it is done with a crowd. The ops staff and managers might see something different. They might throw something in. Sometimes, totally left-field …. From a CRM and training point of view, the NPAS needs to involve everybody …. The RAF … used to have a very good film that highlighted to backroom staff why they are part of flight safety. A Harrier pilot had a crash, and the immediate reaction was 'pilot error'. The film went through the pilot's day prior to the crash. The chef had turned up late, so the pilot did not get his breakfast. Just as he got into work, he got a'phone call from the Pay Office saying they had screwed up his pay, and there was nothing on his pay slip. He had to see his Commander, who told him that a party he was organising had been brought forward. All piling on the stress. Then he goes flying …. Everyone in the organisation needs to be aware of the impact they can have through something they have done or something they have not done". The accident described in the RAF training film was an *organisational* accident (see Reason's 1997 definition).

2.9 Education Would Improve the Working Relationship with Ground Units

Some claimed that education of force police officers in the capabilities of the helicopter would improve operational effectiveness. A TFO explained: "There is a lack of understanding as to what is involved in our role …. We have asked whether we can talk to the new recruits. We've been told there's no time within the timetable …. It's ridiculous, because the helicopter is a fantastic resource … which is a really expensive resource if it is tasked incorrectly".

2.10 Debriefs Are Not Always Performed When Needed

Some interviewees claimed debriefs were routinely performed, others that they were not. A TFO explained: "What you tend to do is, if any one member of the crew feels there is an issue, then they'll say: 'Do you want to have a little chat over a coffee, Chaps, about that last one?' Normally, someone will have a point to make. If you have new crew you will debrief every flight as a matter of course. If you are on an assessment, you will be debriefed as part of your assessment. Maybe debriefing should be compulsory. Maybe you should add it to the routine. It would only take five minutes". Interviewees' different understandings of the status of the post-sortie debrief supports a Base Commander-TFO's observation that each NPAS base possesses a unique *modus operandi*: "[T]he NPAS is a national organisation – has been for four years – and we still have not got set SOPs that are national. Every base does things in different ways. Which is wrong. There should be interoperability". This assessment resonates with Eyre's (Eyre 1984) observations on organisational sub-cultures.

2.11 *Crewmember Ad-Hoc Practices Support Teamworking*

Several claimed that team function was helped by crewmember-initiated work-arounds. A pilot explained: "After you have done all your training, there are certain things that you don't know. For example, there are six police and two air traffic channels to be monitored. In an ideal world, all crew would monitor all channels all of the time. You can't do that. You'd max-out [become task-saturated]. I listen to the air traffic channel and turn down all the other channels. I ask the TFOs to tell me if they are given new operational orders or requests, because I have turned those channels down. Then I can concentrate. Conversely, the TFOs want the tactical information. They turn the air traffic channels down and turn the other channels up. At some point, once we are safely in flight, we can start putting the volumes up on the radios. Nobody ever teaches you that. It's not a SOP. That's pure experience that tells me that's the best way to do it. Little things like that should be filtered into the CRM training". Snook (Snook 2000) describes how over-extended and/or under-resourced personnel may modify procedures, or circumvent them altogether, to meet production targets.

3 Conclusions

There are pressures without and within the NPAS. The government's austerity programme has seen police budgets cut. The police are expected to do more with less. Closure of NPAS bases means that helicopters have to travel further to jobs. The small EC135s are being pushed to their limits. In an ideal world, the aging EC135s would be replaced with larger, more capable, longer-ranged helicopters. As it is, the NPAS finds itself trapped in a perfect storm of budget cuts, high expectations and negative publicity (Bennett 2018).

Official audits paint a bleak picture. In November, 2017, Her Majesty's Inspectorate of Constabulary and Fire and Rescue Services (HMICFRS) published a comprehensive review of the NPAS. The HMICFRS review asserted that the NPAS was failing in its mission: "[W]e found some fundamental problems with the current collaborative arrangements for police air support, which have led us to conclude that urgent change is required – if not a fresh approach entirely …. With the number of bases being halved and the number of aircraft being cut by a third in the last 10 years, savings have primarily been made by cutting the service provided to forces, rather than increasing efficiency. An inconsistent service means that many incidents requiring air support are over before a police helicopter can arrive. Moreover, we are concerned that the police service now operates insufficient aircraft to provide a consistently prompt response to incidents in all forces in England and Wales" (Her Majesty's Inspectorate of Constabulary and Fire and Rescue Services 2017).

Pressures within the NPAS include: tensions between the bases and Headquarters; limited understanding amongst Dispatchers and serving officers of police helicopter capabilities; operation of an aging fleet; providing cover to a growing population with fewer machines; being targeted by shooters; operating a physically and psychologically demanding four-days-on, four-days-off roster; irregular meals; and poor-quality or inadequate sleep. It cannot be claimed that the NPAS is perfect: some aspects of the service's work could be improved. For example, a more consistent approach to the post-sortie debrief would generate a higher-fidelity picture of missions and performance, creating more opportunities for active learning (Toft 1992; Toft and Reynolds 1997) and double-loop learning (Argyris 1976). Nevertheless, despite the pressures, deprivations, threats and occasional failures, the testimony presented here attests to the resilience, dedication and ingenuity of NPAS pilots and TFOs.

The importance of teamwork-training to performance was a key finding. Teamwork-training, known in aviation circles as crew resource management, was universally praised. Some claimed their CRM training helped them accommodate stressors that, *absentia* CRM training, might have caused them to panic or freeze. Others, that their CRM training had helped them grow as individuals. The correlation between teamwork-training and performance makes the regular police service's eschewing of such training all the more puzzling. It is recommended that the police service considers, as a matter of urgency, the introduction of teamwork training for officers (of all ranks). In all probability, the benefits in terms of co-ordination and optimal use of scarce resources would outweigh the costs.

Successive governments have starved the police service of resources, preferring to invest in Britain's National Health Service, a dysfunctional, inefficient and self-serving organisation that kills patients and bullies or fires courageous, well-intentioned whistle-blowers (Francis 2013; Bennett 2017, 2019b; Gosport Independent Panel 2018). The NHS's organisational culture could reasonably be described as pathogenic. Given that the state's primary mission is to ensure the safety and security of its citizens, it is time that spending priorities were re-ordered in favour of policing and national defence. The dedicated men and women of the NPAS, and their hard-pressed colleagues on the ground, deserve nothing less.

Appendix: Interview Questions to: NPAS Pilots and Tactical Flight Officers

1. In what ways has your CRM training helped in your routine, day-to-day work?

2. In what ways has your CRM training helped in non-normal situations?
3. What changes would you make to CRM training to make it more relevant to your role in the cockpit/in the cabin?
4. Do you think CRM could be used in other professions/industries?
5. If 'yes', which professions/industries, and what benefits might CRM bring?
6. Is there anything else you would like to say?

References

Argyris, C. (1976). *Increasing leadership effectiveness*. New York: Wiley.

Armstrong, M. (1996). *A handbook of personnel management practice*. London: Kogan Page.

Barrett, D. H., Resnick, H. S., Foy, D. W., Dansky, B. S., Flanders, W. D., & Stroup, N. E. (1996). Combat exposure and adult psychosocial adjustment among US Army veterans serving in Vietnam, 1965–1971. *Journal of Abnormal Psychology, 105*(4), 575–581.

Bennett, S. A. (2010). A longitudinal ethnographic study of night freight pilots. *Journal of Risk Research, 13*(6), 701–730.

Bennett, S. A. (2017). Aviation's Normal Operations Safety Audit: A safety management and educational tool for health care? *Results of a small-scale trial, Risk Management and Healthcare Policy, 10*, 147–165.

Bennett, S, A. (2018, February). National Police Air Service. *Air International*, 64–67.

Bennett, S. A. (2019a, February). Managing resources. *Air International*, 76–81.

Bennett, S. A. (2019b). The 2018 Gosport Independent Panel report into deaths at the National Health Service's Gosport War Memorial Hospital. Does the culture of the medical profession influence health outcomes? *Journal of Risk Research*. https://doi.org/10.1080/13669877.2019.1591488.

Braun, V., & Clarke, V. (2006). Using thematic analysis in psychology. *Qualitative Research in Psychology, 3*(2), 77–101.

British Broadcasting Corporation. (2015). *Nicholas Salvador detained over woman's beheading*. Available at: https://www.bbc.co.uk/news/uk-england-33239870/. Accessed 15 Feb 2019.

Brookes, A. (1992). *Disaster in the air*. Shepperton: Ian Allan.

Brown, A. D. (1995). *Organisational culture*. London: Pitman Publishing.

Campbell, R. D., & Bagshaw, M. (1999). *Human performance and limitations in aviation*. Oxford: Blackwell Science.

Card, J. J. (1987). Epidemiology of PTSD in a national cohort of Vietnam veterans. *Journal of Clinical Psychology, 43*(I), 6–17.

Christie, D. J., & Evans-Pim, J. (Eds.). (2012). *Nonkilling psychology*. Honolulu: Centre for Global Nonkilling.

Civil Aviation Authority. (2016a). *CAP 737. Flight-crew human factors handbook*. London: Civil Aviation Authority.

Civil Aviation Authority. (2016b). *Standards Document No. 29, Version 7. Guidance on the requirements that pertain to Flightcrew for the training and testing of Human Factors under EASA Part – ORO and EASA Part – FCL*. London: Civil Aviation Authority.

Clay-Williams, R., Greenfield, D., Stone, J., & Braithwaite, J. (2014). On a wing and a prayer: An assessment of modularised crew resource management training for health care professionals. *Journal of Continuing Education in the Health Professions, 34*(1), 56–67.

Collins, D. (2017). *Statement on HMICFRS report into the National Police Air Service*. Available at: https://www.westyorkshire-pcc.gov.uk/news-events/archived-news/archived-news-2017/2017/statement-on-hmicfrs-report-into-the-national-police-air-service-(npas)-(30-nov-2017).aspx/. Accessed 1 Feb 2019.

Davis, M. C., Challenger, R., Jayewardene, D. N. W., & Clegg, C. W. (2014). Advancing socio-technical systems thinking: A call for bravery. *Applied Ergonomics, 45*(2A), 171–180.

Dixon, B. R., Bouma, G. D., & Atkinson, G. B. J. (1987). *A handbook of social science research: A comprehensive and practical guide for students*. Oxford: Oxford University Press.

Eyre, E. C. (1984). *Mastering basic management*. London: Macmillan Education.

Federal Aviation Administration. (2017). Avianca Airlines B707 Flight 52 at Long Island, NY. Lessons Learned. Available at: http://lessonslearned.faa.gov/ll_main.cfm?TabID=4&LLID=70&LLTypeID=2/. Accessed 9 Oct 2017.

Francis, R. (2013). *Report of the mid Staffordshire NHS foundation trust public inquiry. Executive summary*. London: Her Majesty's Stationery Office.

Gordon, S., Mendenhall, P., & O'Connor, B. B. (2013). *Beyond the checklist. What else healthcare can learn from aviation teamwork and safety*. Ithaca: ILR Press.

Gosport Independent Panel. (2018). *Gosport War Memorial Hospital. The report of the Gosport Independent Panel*. London: Her Majesty's Stationery Office.

Green, R., Muir, H., James, M., Gradwell, D., & Green, R. L. (1996). *Human factors for pilots*. Aldershot: Ashgate Publishing Ltd..

Hatch, M. J. (1997). *Organisation theory: Modern, symbolic and postmodern perspectives*. Oxford: Oxford University Press.

Her Majesty's Inspectorate of Constabulary and Fire and Rescue Services. (2017). *Planes, drones and helicopters. An independent study of police air support*. London: Her Majesty's Inspectorate of Constabulary and Fire and Rescue Services.

Hodgkinson, P. E., & Stewart, M. (1991). *Coping with catastrophe: A handbook of disaster management*. London: Routledge.

House of Commons Home Affairs Committee. (2018). *Policing for the future*. London: House of Commons Home Affairs Committee.

Kanki, B. G., Anca, J., & Chidester, T. R. (Eds.). (2019). *Crew resource management* (3rd ed.). San Diego: Academic Press.

Kilmann, R. H. (1989). *Managing beyond the quick fix*. San Francisco: Jossey-Bass.

Krause, S. S. (1996). *Aircraft safety. Accident investigations, analyses and applications*. New York: McGraw-Hill.

Langer, E. J. (1989). Minding matters: The consequences of mindlessness-mindfulness. In *Advances in experimental social psychology* (pp. 137–173). New York: Academic Press.

MacLeod, N. (2005). *Building safe systems in aviation. A CRM developer's handbook*. Aldershot: Ashgate Publishing Ltd..

MacNair, R. (2012). Psychology of nonkilling. In *Nonkilling psychology* (pp. 111–131). Honolulu: Centre for Global Nonkilling.

Mason, R. O. (2004). Lessons in organisational ethics from the Columbia disaster: Can a culture be lethal? *Organisational Dynamics, 33*(2), 128–142.

Maurino, D. E., Reason, J., Johnston, N., & Lee, R. B. (1998). *Beyond aviation human factors – Safety in high-technology systems*. Aldershot: Ashgate Publishing Ltd..

Maxfield, M. G., & Babbie, E. (2001). *Research methods for criminal justice and criminology* (3rd ed.). Boston: Wadsworth Thomson Learning.

McAllister, B. (1997). *Crew resource management. Awareness, cockpit efficiency and safety*. Shrewsbury: Airlife.

Monarch Airlines. (2017). *Cabin Crew Initial CRM* (CRM training course PowerPoint). Monarch Airlines, London Luton Airport

National Police Air Service. (2019). *National Police Air Service*. Available at: https://www.npas.police.uk/. Accessed 1 Feb 2019.

Neuman, W. L. (1997). *Social research methods. Qualitative and quantitative approaches* (Vol. 3). Boston: Allyn and Bacon.

Phillimore, J., Bradby, H., Knecht, M., Padilla, B., & Pemberton, S. (2018, June). Bricolage as a conceptual tool for understanding access to healthcare in superdiverse populations. *Social Theory and Health, 26*, 1–22. https://doi.org/10.1057/s41285-018-0075-4.

Reason, J. T. (1990). *Human Error*. Cambridge: Cambridge University Press.
Reason, J. T. (1997). *Managing the risks of organisational accidents*. Aldershot: Ashgate Publishing Ltd..
Reason, J. (2013). *A life in error*. Farnham: Ashgate.
Sanchez-Burks, J., Karlesky, M. J., & Lee, F. (2015). Psychological bricolage: Integrating social identities to produce creative solutions. In *The Oxford handbook of creativity, innovation and entrepreneurship* (pp. 93–102). New York: Oxford University Press.
Snook, S. (2000). *Friendly fire: The accidental shootdown of U.S. Black Hawks over Northern Iraq*. Princeton: Princeton University Press.
Taylor, P., Richardson, J., Yeo, A., Marsh, I., Trobe, K., & Pilkington, A. (1995). *Sociology in focus*. Ormskirk: Causeway Press.
Thompson cited in Dodd, V. (2018). Policing at 'tipping point' over budget cuts, warns police chief. *The Guardian*, 10 October.
Toft, B. (1992). The failure of hindsight. *Disaster Prevention and Management: An International Journal, 1*(3), 48–63.
Toft, B., & Reynolds, S. (1997). *Learning from disasters*. Leicester: Perpetuity Press.
Trice, H. M. (1993). *Occupational subcultures in the workplace*. New York: ELR Press.
Trist, E., & Bamforth, K. (1951). Some social and psychological consequences of the longwall method of coal-getting. *Human Relations, 4*(1), 3–38.
Tuckman, B. W., & Jensen, M. C. (1977). Stages of small group development revisited. *Group and Organisational Studies, 2*, 419–427.
Turner, B. A. (1988). Connoisseurship in the study of organisational cultures. In *Doing research in organisations*. London: Routledge.
Waring, A. E. (1996). *Safety management systems*. London: Chapman and Hall/ITP.
Watson, T. J. (2003). *Sociology, work and industry* (4th ed.). London: Routledge.
Wiener, E. L., Kanki, B. G., & Helmreich, R. L. (Eds.). (1993). *Cockpit resource management*. San Diego: Academic Press.
Wikimedia Commons. (2011). *Shop fire during London riots*. Available at: https://commons.wikimedia.org/wiki/File:Shop_fire_during_London_riots,_2011.jpg/. Accessed 13 May 2019.
Wikimedia Commons. (2016). *G-NWOI Eurocopter EC135 Helicopter National Police Air Service*. Available at: https://commons.wikimedia.org/wiki/File:G-NWOI_Eurocopter_EC135_Helicopter_National_Police_Air_Service_(29494322826).jpg/. Accessed 13 May 2019.
Wikler, N. (1990). Hidden injuries of war. In *Strangers at home: Vietnam veterans since the war* (pp. 87–106). New York: Brunner/Mazel Inc.

Science Informed Major Event Security Planning: From Vulnerability Analysis to Security Design

Anthony J. Masys

Abstract The post 9/11 security landscape has seen a fundamental shift in security planning, organization and management across the national and global security domain. This shift is particularly evident with regards to special events. Major national regional and global events such as the Olympic Games, FIFA world cup, the US Super Bowl, and the Boston Marathon draw large crowds of spectators and participants thereby creating vulnerable targets. The 1972 Munich Olympics and 1996 Atlanta Olympics are 2 examples of pre-9/11 attacks on sporting events. Since 9/11, we have seen attacks on events such as the Boston Marathon in 2013, Las Vegas active shooter in 2017 and the bombing in Manchester at the Ariana Grande concert in 2017. These examples highlight the inherent vulnerabilities and security challenges that reside in major events (sporting, cultural, religious, musical). With this in mind, security budgets have increased to deal with such risks and threats. For example, the security budget for the 2012 London Olympics topped $1.46 Billion and the 2010 Vancouver Olympics was over $900 million.

This chapter examines the security challenges and solutions for managing major event security. Scenario planning and disaster forensic methodologies (Masys AJ (ed), Disaster forensics: understanding root cause and complex causality. Springer, Cham, 2016) are presented along with the role of predictive analytics in supporting vulnerability analysis and security solution navigation. Key recommendations are made pertaining to vulnerability analysis and the development and design of a security architecture planning framework rooted in a High Reliability Security Organizations (Masys AJ, Designing high reliability security organizations for the homeland security enterprise. In: A Siedschlag, A Jerkovic (eds) Homeland security cultures: enhancing values while fostering resilience. Rowman & Littlefield International, London).

Keywords Major event · Security planning · Vulnerability · Scenario planning · Network analysis

---

A. J. Masys (✉)
College of Public Health, University of South Florida, Tampa, FL, USA
e-mail: tmasys@usf.edu

B. Fox et al. (eds.), *Science Informed Policing*, Advanced Sciences and Technologies for Security Applications,
https://doi.org/10.1007/978-3-030-41287-6_12

1 Introduction

Major events are often valuable opportunities for local governments by providing increased revenue, revitalizing a city, and creating a sense of community. To enable these opportunities to emerge, security and safety for the public must be ensured. The post 9/11 security landscape has seen a fundamental shift in security planning, organization and management across the national and global security domain. This shift is particularly evident with regards to special events. Major national regional and global events such as the Olympic Games, FIFA world cup, the US Super Bowl, and the Boston Marathon draw large crowds of spectators and participants thereby creating vulnerable targets. The 1972 Munich Olympics and 1996 Atlanta Olympics are 2 examples of pre-9/11 attacks on sporting events. Since 9/11 we have seen attacks on events such as the Boston Marathon in 2013, Las Vegas active shooter in 2017 and the bombing in Manchester at the Ariana Grande concert in 2017. These examples highlight the inherent vulnerabilities and security challenges that reside in major events (sporting, cultural, religious, musical). With this in mind security budgets have increased to deal with such risks and threats. For example, the security budget for the 2012 London Olympics topped $1.46 Billion and the 2010 Vancouver Olympics was over $900 million.

The London Olympic Strategic Threat Assessment (OSTA) assessed the threats and hazards associated with the safety and security of the Games. These fell into the following categories (Home Office 2011):

- terrorism
- serious crime
- domestic extremism and public disorder
- natural hazards.

As articulated by Taylor and Toohey (2017:379), in the post 9/11 era '…security budgets have escalated in response to contemporary perceptions of threats of terrorism. Although there is a low risk of an actual terrorism incident occurring during an event, there appears to be heightened public demand for all possible contingencies to be covered. While tens of thousands of sports events run each year without incident, it only takes one high profile situation, such as the Boston Marathon bombings on 15 April 2013, which killed three people and injured 264 others, to increase public demand for 'better' security and safety'.

The link between terrorism and sport is not a new development (it was most dramatically evident at the Olympics of 1972 and 1996). Table 1 presents highlights of some post 9/11 examples of such events.

Security risks vary depending on the size, type, and context of the major event. Taylor and Toohey (2017:374) note that generally, the majority of risks are associated with illegal crowd behaviour (e.g. public intoxication, spectator aggression), local crime (e.g. ticket scalping), political activism (e.g., 'anti-event' protests, strikes), and terrorism (e.g. bombs or shootings). With regards to organized crime and major events, Fussey et al. (2012) describe the experience of the Atlanta, Sydney

Table 1 Major event security occurrences

2002	ETA detonated a car bomb outside the Bernabau stadium just before a Champions League football match between Real Madrid and Barcelona
2008	A suicide bomber attacked the start of a marathon outside Colombo, killing 12 people including a former Olympic marathon runner and a national athletics coach
2008	It was reported that the Dakar rally was cancelled because of 'direct' threats of terrorism (Guardian, January 2008)
2009	Six policemen and a bus driver were killed and seven cricketers and a coach were injured when the Sri Lankan cricket team's convoy was attacked by gunmen in Lahore
2010	Nearly 100 people were killed when a suicide bomber attacked a volleyball match in Laki Marwat, Pakistan
2013	Boston Marathon terrorist attack
2017	Las Vegas active shooter at Route 91 Harvest music festival
2017	Bombing at the Arianna Grande concert in Manchester, UK
2019	An active shooter incident took place at the Garlic Festival in Gilroy, CA. The Garlic Festival is an annual event which annually entertains 80,000 to 100,000 people

Richards (2012)

and Athens Summer Olympiads where the proliferation of the sex industry and in particular human trafficking was seen. Given the inherent complexity associated with security planning with consideration of local and organized crime and the emergence of black swan events (Taleb 2007), the organization and delivery of sport events now requires an increasingly systematic approach to governance, including planning and managing for every kind of possible threat and risk.

2 Major Event Security Planning: Exploring Scenarios

Taylor and Toohey (2017:373) argue that Security at sports events is affected by a range of factors, from local contextual and historical aspects (e.g. spectator related tensions) to larger geopolitical influences (e.g. political or religious motivation). The complexity associated with large scale events is rooted in a number of inherent characteristics illustrated in Fig. 1. All these characteristics shape and are shaped by the threat and hazard risk assessment and thereby influence the vulnerability and security analysis and planning.

The requirement for threat, hazard and vulnerability analysis is thereby key to security planning. A new paradigm and methodologies have emerged to support security planning and operations. A focus on public safety, security, and risk mitigation has precipitated the requirement for enhanced capabilities to support risk management including innovation and improvements in event security management technology, staff training, and operational procedures and practices. Recognizing that nothing should go wrong opens up a reflective practice that explores what can go wrong thereby emphasizing the requirements for collaborative security planning amongst the various stakeholders associated with the major event.

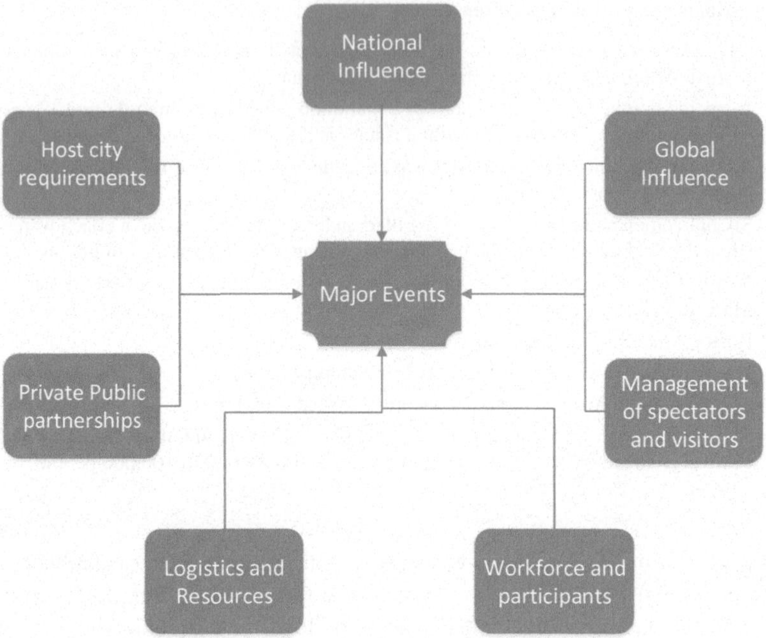

Fig. 1 Large scale event characteristics that influence the security planning

The All Hazards Risk Taxonomy (https://www.publicsafety.gc.ca/cnt/rsrcs/pblctns/ll-hzrds-ssssmnt/ll-hzrds-ssssmnt-eng.pdf), Fig. 2, provides an organized schema categorizing threats along man-made intentional, man-made unintentional and natural disasters. This taxonomy can be used to support scenario planning and contextualize threats and hazards for specific major events.

3 Scenario Planning

In support of major event security and safety planning, the use of scenarios stretches and refocuses thinking and mental models thereby challenging the planning assumptions to recognize the fundamental uncertainty associated with the emergence of black swans. With this in mind, the scenario planning process supports "thinking the unthinkable" exploring vulnerabilities and uncertainty to recognize alternate futures in a space of possibilities. Masys (2012) describes a variety of scenario planning approaches to help shed light on the existence of black swans.

As described in Masys (2012),

> Black swans represent the unpredictable. They represent '… our misunderstanding of the likelihood of surprises' (Taleb 2007: 2). A black swan is described by Taleb (2007) as that which is an outlier, that which is outside the realm of regular expectations which carries with it an extreme impact such as natural disasters, market crashes, catastrophic failure of

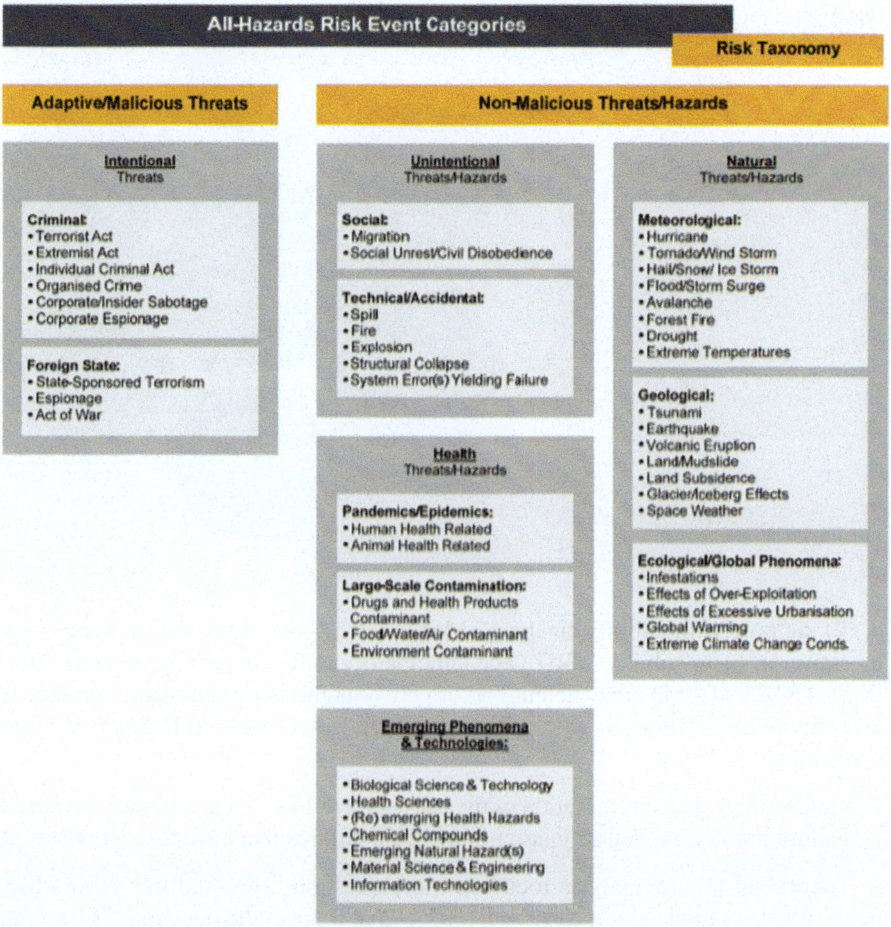

Fig. 2 All hazards risk taxonomy (https://www.publicsafety.gc.ca/cnt/rsrcs/pblctns/ll-hzrds-ssss-mnt/ll-hzrds-ssssmnt-eng.pdf)

complex socio-technical systems and terrorist events such as 9/11. These "surprising events" reflect an organizations inability to recognize evidence of new vulnerabilities or the existence of ineffective countermeasures (Woods 2006:24). This necessitates the requirement to readjust to their existence and thereby the need to consider the extremes (Taleb (2007: xx).

Exploring the black swan plausibility space is rooted in a scenario analysis methodology (Scenario Planning^{XT}), as shown in Fig. 3, that encompasses a number of tools, mindsets and approaches such as: Concept mapping, AHRA taxonomy, Anticipatory Failure Determination, Red Teaming, Network Analysis, Systems Thinking. These are described in Masys (2012, 2015) and Xu and Masys (2016).

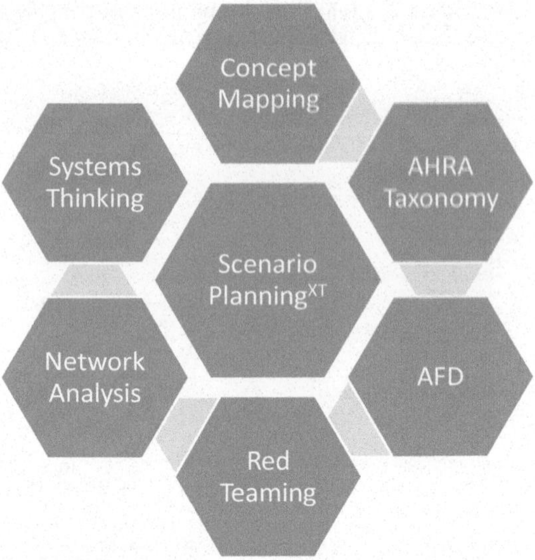

Fig. 3 Scenario planning^{XT}

To support vulnerability analysis, Masys (2012) describes the application of inventive problem solving (TRIZ) and anticipatory failure determination (AFD). In short, TRIZ and AFD are efficient and effective methods for analyzing, predicting and eliminating failures in systems, products and processes. AFD has two broad applications:

1. Failure analysis: determination of the cause of a failure that has already occurred.
2. Failure prediction: determination of possible failures that have not yet occurred.

Traditional failure analysis focuses on the question "How did this failure happen?" What is unique about AFD and TRIZ is that it poses the question "If I wanted to create this particular failure, how could I do it?" Failure prediction thereby reflects an iterative application of failure determination to envision all the possible end states, mid states, initiating events and possible scenarios leading to these states thereby revealing interdependencies and black swans. This becomes a powerful approach to support security vulnerability analysis.

As described in Masys (2012), the process of red teaming provides a powerful methodology to serve as a devil's advocate, offering alternative interpretations and challenging established thinking in order to identify and reduce risks. Within the military domain in particular, red teaming has an outstanding record of success. Within this domain and others, red teaming is used as a way to challenge planning assumptions, established mental models, feasibility, vulnerability and risk and thereby helping to avoid the common fault of group think (Janis 1972). As described in Masys (2012) '…Within the security domain such as preparation for an international major event (London 2012 Olympics, Brazil 2014 FIFA World Cup and 2016

Olympics), red teaming can be used to conduct peer review of plans and policies to detect vulnerabilities and present alternative scenarios to challenge the problem space. Applied to the domain of security, Mateski (2009: 35) reports that '... red teaming is considered the best tool for raising security awareness in an organization and as such can facilitate the discovery of security holes for which known fixes, configurations, or patches have not been applied or where compensating security procedures are not in effect or not being enforced'.

The tools, mindsets and approaches of scenario planning[XT] challenge the assumptions and mental models that shape risk, crisis and disaster management. As revealed from the case studies described in Masys (2012), scenario planning and worst-case scenarios are key to explore the boundaries of uncertainty to facilitate preparedness in support of risk, crisis and disaster management. Taleb (2007) argues that "...what is surprising is not the magnitude of our forecast errors, but our absence of awareness of it .... Owing to this misunderstanding of the causal chains between policy and actions, we can easily trigger Black swans thanks to aggressive ignorance". The case studies described in Masys (2012) highlight how understanding these interdependencies through a systems thinking lens and leveraging network analysis to support cascading disasters are essential to foresight.

As described by Dekker (2011:97), 'That which cannot be believed will not be seen'. Scenario planning does not focus on accurately predicting the future (Hodgkinson and Wright 2002) but rather is a process that produces a number of plausible futures that are credible and yet uncertain and thereby reveal black swans that could threaten safety and security. Combining scenario planning and predictive analytics, as described in Stanley et al. (2019) and Gardner et al. (2017), supports an experimental furturing framework that informs public safety and security strategy development and strategic interventions. Through predictive analytics, enhanced vulnerability analysis can be conducted. Scenarios emerging from this are influential across the disaster and security management domain (mitigation, preparedness, response and recovery). As such this takes experimental futuring to a new level of impact in effect, linking scenario planning and predictive analytics with strategy development.

In support of the scenario planning effort, through the application of concept mapping, goals threats and vulnerabilities can be decomposed creating a relational network map of cascading events and impacts. Concept mapping provides a visualization and thought process to support security vulnerability analysis and security design.

A concept map is a diagram or graphical tool that visually represents relationships between concepts and ideas. Most concept maps depict ideas through a structured hierarchically and connected framework. In this way the concept map provides a logical and readily evidence-supported view pertaining to the decomposition and analysis of the threat and hazard thereby identifying intervention strategies and leverage points for security design.

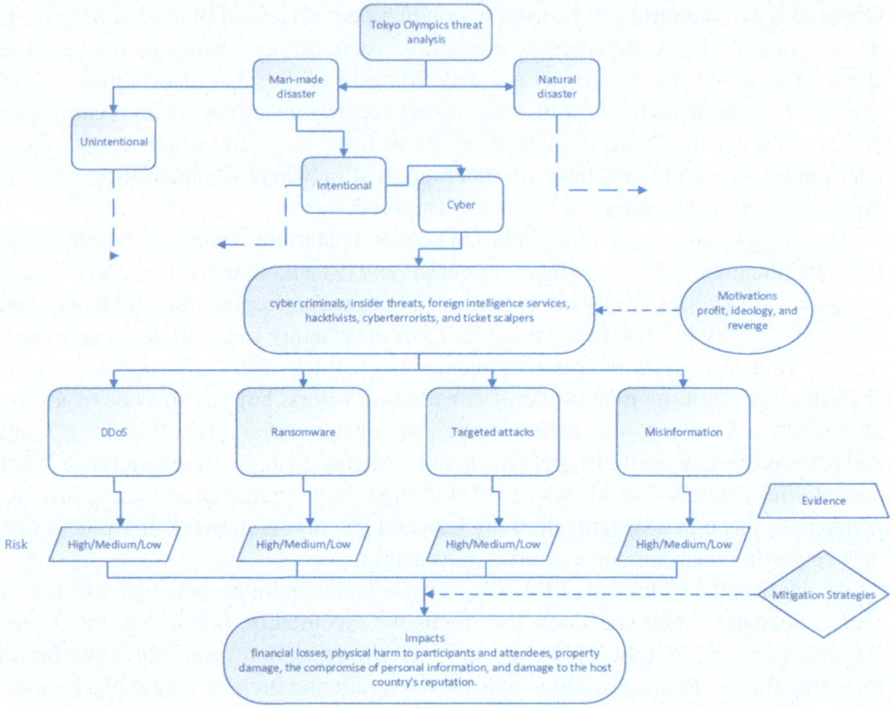

Fig. 4 Concept map of Tokyo 2020 Olympics cyber threat analysis

For example, Fig. 4 depicts a concept map of Cyber threats (RAND 2018) and vulnerabilities associated with the Tokyo 2020 Olympics. Note only the cyber branch of the concept map is shown. A complete map would also depict the natural disasters, unintentional accidents and a full decomposition of all man-made intentional threats as well as depict the influence and interconnectivity that exists across and between the threats and hazards.

As a risk analysis tool as well as a security planning tool, we can apply the concept map of threat, hazards and vulnerability along with the core operational areas associated with security planning (Table 2) to develop mitigation and preparedness strategies.

The key message is that with forward planning, such as risk identification, mitigation strategies, and pre-disaster emergency and recovery planning one can better manage incidents arising from natural and man-made threats and hazards. As such scenario planning and red teaming captures a whole range of possibilities in rich detail to support risk management and security planning.

Table 2 Core operational areas

1. Administrative and logistics support
2. Command and control
3. Credentialing
4. Crowd management
5. Dignitary/VIP protection
6. Financial management
7. Fire/emergency medical services (EMS)/ hospitals/public health
8. Intelligence/counterterrorism/counter surveillance
9. Interagency communications and technology
10. Legal affairs
11. Non-event patrol
12. Prisoner processing
13. Protecting critical infrastructure and utilities
14. Public information and media relations
15. Screening and physical security
16. Tactical support and explosive device response/hazardous materials
17. Training
18. Transportation and traffic
20. Project management
21. Scientific advising
22. Exercise development and evaluation

4 Understanding the Emergence of Disasters (Intrinsic to Security Plans)

Security disaster may emerge not only from external threats but also from the seeds of the disaster rooted within the very security protocols designed. Two theoretical frameworks can be applied to view these unintentional pathways to disaster: Perrow's Normal Accident Theory (NAT) (Perrow 1984), Reason's Swiss Cheese model (2000).

5 Normal Accident Theory

Within the context of major event security, it is recognized that security planning is complex and operates in VUCA conditions (volatility, uncertainty, complexity, ambiguity). As such security planning encompasses a complex and interconnected web of organization, operations and technology that makes major events vulnerable to the 'normal accidents'. Perrow (1999:12) remarks that 'we have produced designs so complicated that we cannot anticipate all the possible interactions of the inevitable failures; we add safety devices that are deceived or avoided or defeated by hidden paths in the systems. The systems have become more complicated because either they are dealing with more deadly substances, or we demand they function in ever more hostile environments or with ever greater speed and volume'. What characterizes "normal accidents" as an inevitable event are the precursors of what Perrow (1984) calls "interactive complexity" and "tight coupling". Tight coupling refers to a system in which as Perrow (1984:4) argues '…processes happen very fast and can't be turned off; the failed parts cannot be isolated from the other parts'. Interactive complexity refers to a system design '…so complicated that we cannot anticipate all the possible interactions of the inevitable failure' (Perrow 1984:11).

Contextualized for security planning, no matter how effective security protocols and procedures are, there is a form of 'accident' that is inevitable arising from the way failures can interact and the way the system is tied together. In other words, the interactions between human and technical systems also constitute systems and these systems represent vital blind spots which are often opaque giving rise to 'surprises'.

Jennings (2012) provides a detailed discussion regarding matters pertaining to Optimism bias, discounting, risk-aversion and normal accidents. It is well recognized that with the inherent complexity associated with security planning, bespoke design of security infrastructure and facilities, all delivered just- in-time, limits opportunities for stress- testing and event rehearsals thereby shaping risk perception, risk communication and expectations. Without testing and applying reflective practices, intrinsic security dysfunctionalities may reside and go unnoticed.

6 Swiss Cheese Model (Reason 2000)

As an accident model, the Swiss Cheese Model describes accidents as a combination of specific events and the failure of one or more barriers – or of all barriers. Inherent within security planning there are leadership, organizational and management culture, technical specifications supporting security, team communications, coordination, training, and security procedures. When dysfunctionalities exist across these areas, it can provide a trajectory whereby an accident and disaster emerges (Fig. 5).

The Swiss Cheese Model (Reason 2000) illustrates how failed barriers (intrinsically designed into the security plans) can be found at any level of the organization. With this in mind, the model provides an excellent interrogation of the security plans to identify vulnerabilities, assumptions and expectations that need to be realized in order to design effective security plans.

Applying these accidents models as a lens during the development of security plans facilitates a reflective practice that feeds into red teaming and scenario planning in order to produce reliable and effective plans.

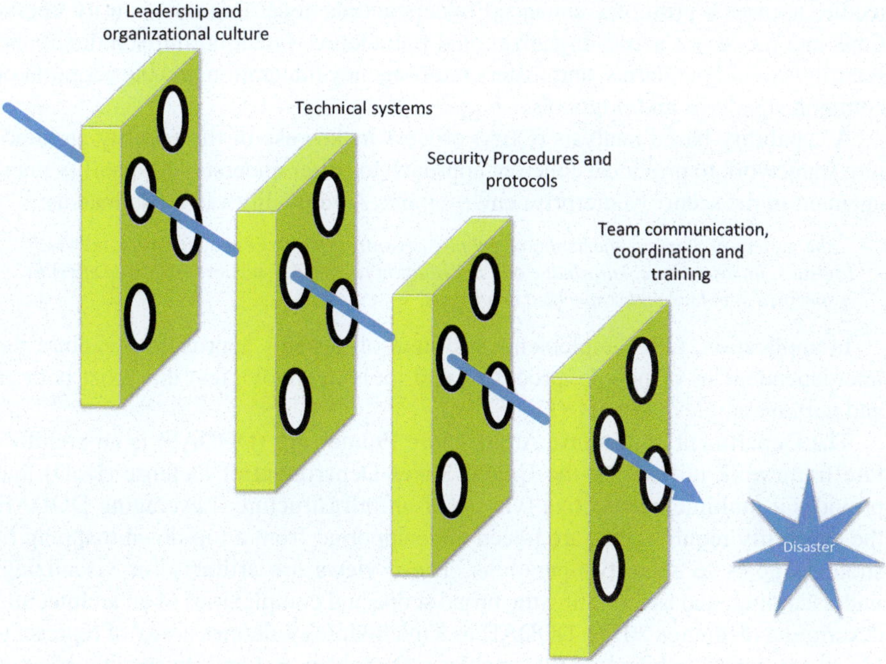

Fig. 5 Swiss cheese model

7 Security Planning: Strategic, Capability and Operational Views

7.1 Architecture Framework Methodology

Scenario planning is at the heart of security as it reveals contextual threats and vulnerabilities thereby supporting the development of a risk and vulnerability analysis to inform security plans. Major event security planning requires a methodology that aligns strategic, operational and tactical planning. A security architecture can be viewed as a set of blueprints which model or represent a wide variety of relationships inherent to the overall security capability being managed. It enables the alignment of people, processes and technologies that exist (the "as is") or need to exist (the "to be") within a capability. The architecture framework outlines "what" the overall structured approach is for assisting interoperability and "how" the components will operate.

In support of security design, strategic, capability and operational views of the security design enable the analysis of interdependencies and interactions between people, processes and technologies within the security system and across systems thereby supporting a capability design and analysis. In so doing, the security architecture approach promotes enhanced understanding regarding the security operations and processes across organizational boundaries, including jurisdictional and security/safety boundaries, and fosters multi-agency integration and the adoption of common standards and protocols.

A capability-based analysis (CBA) process makes use of the security architecture framework to provide a coherent approach to better understand capability management in the security enterprise environment. A capability can be defined as:

> *...the system-of-systems consisting of people, organization, processes, training, materiel, logistics, infrastructure, knowledge and information required to achieve a desired effect to a specified standard under specified conditions.*

In application, CBA implements a system-of-systems approach to capture the interdependencies of people, processes and tools (technologies) that exist both as and within complex systems. (Fig. 6).

The Department of Defense Architecture Framework (DODAF) is an architecture framework utilized by the United States Department of Defense (DoD) that provides a multiple perspective (views) of an infrastructure. Leveraging DODAF, the capability requirements are traced and supported from a top-down mapping of strategic goals to security operations. These views are artifacts for visualizing, understanding, and assimilating the broad scope and complexities of an architecture description (Jennings 2012). DODAF, as a methodology, defines a way of representing an enterprise architecture that enables stakeholders to focus on specific areas of interests in the enterprise while retaining sight of the big picture. The DODAF methodology does this by dividing a problem space into manageable pieces that correspond to stakeholder viewpoints, which are further defined as models. This becomes a powerful methodology for use by various security event planning teams that requires a consolidation of individual plans.

The architectural framework presents data as various views such as: Strategic View (StratV), Capability View (CapV), Operational Views (OV). These views are used to fully document the current state, or the proposed future state, of a system or capability to support security planning (Fig. 7).

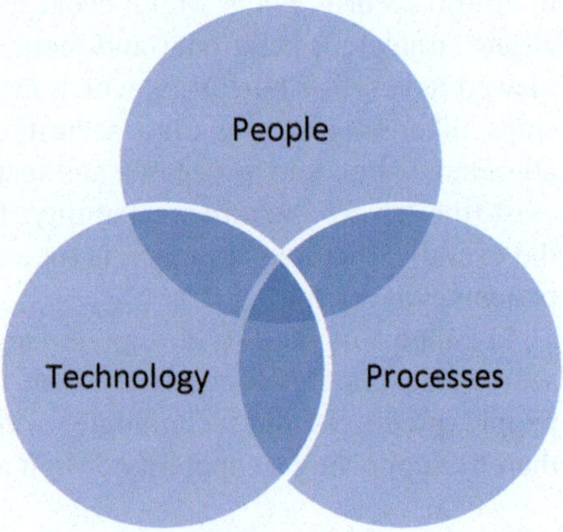

Fig. 6 The people, process and technology provide the means to achieve the capability

Fig. 7 Primary views for security planning

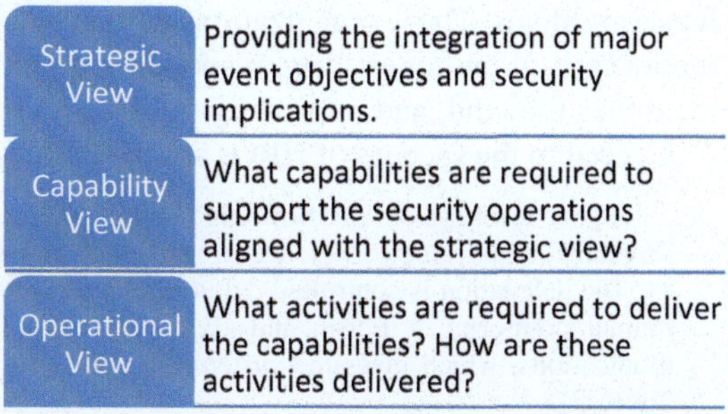

Through this structured approach to security enterprise design, a number of key security requirements emerge such as: roles, responsibilities and relationships; organizational, policy and legislative drivers; operational concepts and critical 'security' processes.

Across these views are core operational areas required to support security planning and operations (Table 2). DODAF representations of each operational area must be consolidated and rolled up into an overall security framework. That will ensure consistency and operational deconfliction across the areas.

8 High Reliability Security Organizations

With the development of a security architecture framework informed by a threat and vulnerability analysis, the establishment of a High Reliability Security Organization (HRSO) (Masys 2018) is an essential element for the conduct of successful security operations. Challenges associated with integrated planning, communications, and interoperability to support major event security operations arise from the inherent complexity of the problem space. The level of effort and the degree of investment in establishing an effective security posture for upcoming major events warrants serious consideration into how new knowledge, practices, and protocols can be captured and institutionalized to support an enhanced degree of local, regional and national resiliency.

The events captured in Table 1 reflect how 'unexpected events often audit our resilience' (Weick and Sutcliffe 2007). Such 'surprising events' often reflect an organizations inability to recognize evidence of new vulnerabilities, sense weak signals. or the existence of ineffective countermeasures (Woods 2006:24). These 'Black swan' events represent the unpredictable and thereby necessitates the requirement to readjust to their existence and the need to consider the extremes (Masys 2018; Taleb 2007:xx).

Successful organizations that can manage complexity and avoid disasters though operating in hazardous and challenging conditions are called High Reliable Organizations (HRO). As described in Masys (2018), the roots of the HRO paradigm finds itself in the examination of aircraft carrier operations, air traffic control and nuclear power operations. They not only avoid dire consequences by catching problems early, but they have actually proven over time to have disproportionately fewer problems. These organizations do this by consistently noticing the unexpected, reporting it in an honest way, responding quickly and appropriately, learning from the things they did, and improving the process for the next time a challenge arises.

Central to the success of HROs are the cultural dimensions. An HRO depends on:

> ...a higher-level system of knowledge and experience, interacting with and supporting an [organizational framework] to transform a high-risk system into a high-reliability system.
>
> The delegation is controlled culturally by a powerful system of selection, training, and mutual monitoring, criticism, and advice. The result is a pattern of extremely efficient communications, which gives the system the ability to absorb damage and surprises (Spann 2006:33).

Weick and Sutcliffe (2007) suggest there are five common concepts that help organizations manage the threat of failure, absorb damage and surprises, and thereby become an HRO. The first three concepts fall under the category of "Anticipate the Unexpected" and the final two are listed under "Contain the Unexpected."

The underlying fabric of a security architectural framework described earlier is the notion of High Reliability Security Organizations (HRSO) (Masys 2018) integrated with the essential property of a sound security culture.

Masys (2018) describes in detail the key requirements of a High Reliability Security Organizations (HRSO), that being Mindset (Fig. 8); Security Culture and HRSO principles. At the core is mindset. How we view the world shapes what we see and then what we do. The complexity associated security planning requires a key foundation in:

- Systems Thinking
  - As described in Masys (2015), in this complex problem landscape, systems thinking is both a worldview and a process in the sense that it informs ones understanding regarding a system and can be used as an approach in problem

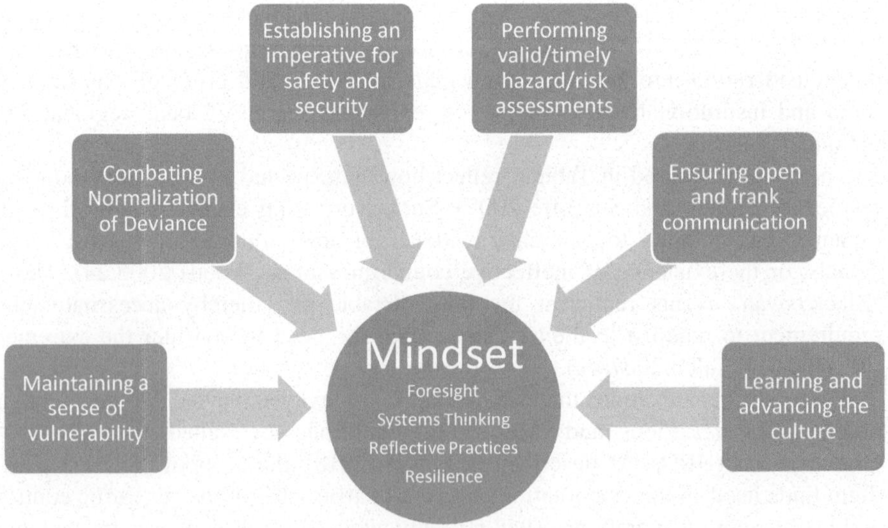

Fig. 8 Lessons learned and mindset. (Masys 2018)

solving (Edson 2008). Senge (1990) describes systems thinking as 'a discipline for seeing wholes…a framework for seeing interrelationships rather than things, for seeing patterns of change rather than static snapshots'. The behaviour of the system is a result of the interaction and interrelationships that exists thereby acknowledging emergent behaviours and unintended consequences. The butterfly defect, as described in Goldin et al. (2014) draws attention to the new nature of systemic risk, which is such that small perturbations now have much greater effects and permeate all dimensions of society.

- Foresight
  - Foresight is the act of looking to and thinking about the future. Foresight enables you to get better at predicting, creating, and leading the future. This is a critical element of scenario planning.
- Reflective Practices
  - Sense making is at the heart of security operations. According to Weick (1995), sense making has a 'strong reflexive quality' to the process of interpretation because 'people make sense of things by seeing a world on which they already imposed what they believe'. In this dynamic and complex risk landscape reflective practices thereby can help to avoid overconfidence through a generative learning approach. Individual reflection is also being a 'reflective practitioner' (Schön 1983) that involves: 'on-the-spot surfacing, criticizing, restructuring, and testing of intuitive understanding of experienced phenomena; often it takes the form of a reflective conversation with the situation' (Schön 1983:241–242).
- Resilience
  - According to Kamara et al. (2019) resilience is broken down to three categories: absorptive, adaptive, and transformative. Absorptive resilience is the ability to take protective action and deal with the stress a disaster brings as it is happening. Adaptive resilience is the ability to prepare and adjust according to a community's current or anticipated situation. Finally, transformative capacity is the ability for a community to identify, minimize, and/or correct risk hazards (Kamara et al. 2019). All three areas must be addressed to minimize impact of disaster on a community.

9 Defining a Security Culture

As described in Masys (2018), security culture emerged from the research in Safety culture identified as being fundamental to an organisation's ability to manage safety-related aspects of its operations. The term security culture has been well articulated in the nuclear industry. Nuclear security culture is defined as:

> The assembly of characteristics, attitudes and behaviour of individuals, organizations and institutions which serves as a means to support and enhance nuclear security. An effective nuclear security culture is dependent on **proper planning, training, awareness, operation** and maintenance, as well as on people who plan, operate and maintain nuclear security systems. ... Ultimately, therefore, the entire nuclear security regime stands or falls because of the people involved and their leaders, and it is the human factor, including management leadership, that must be addressed in any effort to enhance the existing nuclear security culture (IAEA 2008).

Of note is the person-centered vs technology centered approach to security. As described in Masys (2018), the High Reliability Security Organization (HRSO) was designed from an analysis of various major event security operations. Table 3 provides a detailed description of the HRSO.

Table 3 Derived and contextualized from Weick and Sutcliffe (2007) and Masys (2018)

High Reliability Security Operations (HRSO)
HRSO principle 1: Preoccupation with failure HRSO explore the threat and intelligence domain to determine the space of possibilities. They encourage the reporting of errors and miscalculations with regards to security assessments, they elaborate experiences of a near miss for what can be learned, and they are wary of the potential liabilities of success, including complacency, the temptation to reduce margins of security.
HRSO principle 2: Reluctance to simplify HRSOs take deliberate steps to create more complete and nuanced pictures of what they face and who they are as they face it. Knowing that the world they face is complex, stable, unknowable, and unpredictable. HRSOs position themselves to see as much as possible. They welcome diverse experience, skepticism toward received wisdom.
HRSO principle 3: Sensitivity to operations They are attentive to the front line, where the real work gets done. …Anomalies are noticed while they are still tractable and can still be isolated. All of this is made possible because HRSOs are aware of the close ties between sensitivity to operations and sensitivity to relationships.
HRSO Principle 4: Commitment to Resilience The essence of resilience is therefore the intrinsic ability of an organization (system) to maintain or regain a dynamically stable state, which allows it to continue operations after a major mishap and/or in the presence of a continuous stress. The hallmark of HRSO is not that it is incident-free but that incidents don't disable it. Resilience demands a deep knowledge of the threat and space of possibilities (both in time and space).
HRSO Principle 5: Deference to Expertise HRSOs cultivate diversity, not just because it helps them notice more in complex security environments, but also because it helps them do more with the complexities they do spot. Rigid hierarchies have their own special vulnerability to error and misjudgment.
HRSO Principle 6: Adaptability HRSOs embrace a Safety II (Hollnagel 2012) approach that recognizes the need for dynamic and adaptive behavior that is derived from the first 5 principles. HRSOs understand that work as imagined often deviates from work as done. The volatility, uncertainty, complexity and ambiguity that characterize the counter terrorism environment require a certain degree of adaptation to support the very fluid environment. Mindfulness (Weick and Sutcliffe 2007) and reflective practices (Masys et al. 2014) figure prominently.

10 Conclusion

The threat and risk landscape associated with major events is manifest. In fact, between 1972 and 2004, there have been 168 terrorist attacks related to sporting events more generally (Zekulin 2009). Security operations in support of major events operates under VUCA conditions. As such, there is a requirement for a more science informed and methodical approach to security planning. This chapter describes the development of a Security architecture framework that is rooted in scenario planning, DODAF and HRSO. These three elements comprise an approach that characterizes the range of threats and vulnerabilities, design and architecture framework linking strategic, capability and operational views of the security event, and finally an organizational and leadership approach rooted in high reliability for security operations.

References

All Hazards Risk Taxonomy. Available at https://www.publicsafety.gc.ca/cnt/rsrcs/pblctns/ll-hzrds-ssssmnt/ll-hzrds-ssssmnt-eng.pdf

Dekker, S. (2011). *Drift into failure: From hunting broken components to understanding complex systems*. Aldershot: Ashgate Publishing.

DODAF. (2015). *Volume I: Overview and concepts*. Available at: https://dodcio.defense.gov/Portals/0/Documents/DODAF2/DoDAF%20v2.02%20Chg%201%20Vol%20I%20Final%202015-01-19.pdf

Edson, R. (2008). *Systems thinking. Applied: A Primer*. ASysT Institute http://www.anser.org/docs/systems_thinking_applied.pdf

Fussey, P., Richards, A., & Silke, A. (2012). Chapter 12 – Critical reflections on securing the Olympics: Conclusions and ways forward. *Routledge Online Studies on the Olympic and Paralympic Games, 1*(41), 227–238.

Gardner, W. L., Panaou, T., Watterson, K., Alao, S., & Jacob, B. G. (2017). Predicting potential human trafficking recruiting geolocations using Eigen decompressed non-zero spatial autocorrelation coefficients and Poissonian heterogeneous gamma distributed variables. *International Journal of Geographic Information System,4*(2). Retrieved from http://acascipub.com/Journals.php

Goldin, I., & Mariathasan, M. (2014). *The butterfly defect- how globalization creates systemic risks, and what to do about it*. Princeton: Princeton University Press.

Hodgkinson, G., & Wright, G. (2002). Confronting strategic inertia in a top management team: Learning from failure. *Organization Studies, 23*(6), 949–978.

Hollnagel, E. (2012). *A tale of two Safeties*. http://www.resilienthealthcare.net/A_tale_of_two_safeties.pdf

Home Office. (2011). *London 2012 Olympic and Paralympic safety and security strategy*. Available at: https://assets.publishing.service.gov.uk/government/uploads/system/uploads/attachment_data/file/97983/olympic-safety-security-strategy.pdf

IAEA. (2008). *Nuclear security culture* (IAEA nuclear security series no. 7). Available at: https://www-pub.iaea.org/MTCD/Publications/PDF/Pub1347_web.pdf

Janis, I. (1972). *Victims of groupthink*. Boston: Houghton-Mifflin.

Jennings, W. (2012). Chapter 8 – Governing the games in an age of uncertainty: The Olympics and organisational responses to risk. *Routledge Online Studies on the Olympic and Paralympic Games, 1*(41), 135–162. https://doi.org/10.4324/9780203835227_chapter_8.
Kamara, J. K., Agho, K., & Renzaho, A. M. N. (2019, March 1). Understanding disaster resilience in communities affected by recurrent drought in Lesotho and Swaziland—A qualitative study. *PLOS One*. Available at: https://journals.plos.org/plosone/article?id=10.1371/journal.pone.0212994
Masys, A. J. (2012). Black swans to grey swans- revealing the uncertainty. *International Journal of Disaster Prevention and Management, 21*(3), 320–335.
Masys, A. J. (Ed.). (2015). *Applications of systems thinking and soft operations research in managing complexity*. Cham: Springer.
Masys, A. J. (Ed.). (2016). *Disaster forensics: Understanding root cause and complex causality*. Cham: Springer.
Masys, A. J. (2018). Designing high reliability security organizations for the homeland security Enterprise. In A. Siedschlag & A. Jerkovic (Eds.), *Homeland security cultures: Enhancing values while fostering resilience*. London/Lanham: Rowman & Littlefield International.
Masys, A. J., Ray-Bennett, N., Shiroshita, H., & Jackson, P. (2014). High impact/low frequency extreme events: Enabling reflection and resilience in a hyper-connected world. 4th International conference on building resilience, 8–11 September 2014, Salford Quays, United Kingdom. *Procedia Economics and Finance, 18*(2014), 772–779.
Mateski, M. (2009). *Red teaming: A short introduction 1.0*. Available at: http://redteamjournal.com/papers/A%20Short%20Introduction%20to%20Red%20Teaming%20(1dot0).pdf. Accessed 4 Sept 2009.
Perrow, C. (1984). *Normal accidents- living with high-risk technologies*. New York: Basic Books.
Perrow, C. (1999). *Normal accidents- living with high-risk technologies*. Princeton: Princeton University Press.
RAND. (2018). *Olympic-caliber cybersecurity*. https://www.rand.org/pubs/research_reports/RR2395.html
Reason, J. (2000). Human error: Models and management. *BMJ (Clinical Research Ed.), 320*(7237), 768–770. https://doi.org/10.1136/bmj.320.7237.768.
Richards, A. (2012). Chapter 2 – Terrorism, the Olympics and sport: Recent events and concerns for the future. *Routledge Online Studies on the Olympic and Paralympic Games, 1*(41), 15–31.
Schön, D. (1983). *The reflective practitioner: How professionals think in action*. London: Temple Smith.
Senge, P. (1990). *The fifth discipline: The art and practice of the learning organization*. New York: Doubleday Currency.
Spann D. (2006). Coping with the unexpected: Integrating HRO and agile. *Cutter IT Journal*. Available at: http://citeseerx.ist.psu.edu/viewdoc/download?doi=10.1.1.606.2955&rep=rep1&type=pdf
Stanely, N., Jacob, B. G., Masys, A. J., Parikh, J., Izurieta, R., Ortiz, M. (2019). Global Health Security and Disaster Forensics: A solution oriented approach to mapping public health vulnerabilities through predictive analytics. In *E-Proceedings of the 38th IAHR World Congress*, September 1–6, 2019, Panama City, Panama.
Taleb, N. N. (2007). *The black swan: The impact of the highly improbable*. London: Penguin Books Ltd.
Taylor, T., & Toohey, K. (2017). The security agencies' perspective. In M. Parent & J.-L. Chappalet (Eds.), *Routledge handbook of sports event management*. London: Routledge.
Weick, K. E. (1995). *Sensemaking in organizations*. Thousand Oakes: SAGE.
Weick, K. E., & Sutcliffe, K. M. (2007). *Managing the unexpected: Resilient performance in an age of uncertainty* (2nd ed.). San Francisco: Wiley.
Woods, D. D. (2006). Essential characteristics of resilience. In E. Hollnagel, D. D. Woods, & N. Leveson (Eds.), *Resilience engineering: Concepts and precepts* (pp. 21–34). Aldershot: Ashgate Publishing.

Xu, T., & Masys, A. J. (2016). Critical infrastructure vulnerabilities: Embracing a network mind-set. In A. J. Masys (Ed.), *Exploring the security landscape- non-traditional security challenges*. Cham: Springer.

Zekulin, M. (2009, Fall) Olympic security: Assessing the risk of terrorism at the 2010 Vancouver Winter Games. *Journal of Military and Strategic Studies*, *12*(1)1–25

Index

B. Fox et al. (eds.), *Science Informed Policing*, Advanced Sciences and Technologies for Security Applications,
https://doi.org/10.1007/978-3-030-41287-6

The manufacturer's authorised representative in the EU is Springer Nature Customer Service Centre GmbH, Europaplatz 3, 69115 Heidelberg, Germany. If you have any concerns regarding our products, please contact ProductSafety@springernature.com

Printed and bound by CPI Group (UK) Ltd, Croydon, CR0 4YY
30/09/2026
02225445-0007